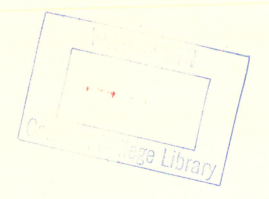

Geologic Maps

A Practical Guide to the Preparation and Interpretation of Geologic Maps

For Geologists, Geographers, Engineers, and Planners

Second Edition

Edgar W. Spencer
Washington and Lee University

PRENTICE HALL
Upper Saddle River, New Jersey 07458

Library of Congress Cataloging-in-Publication Data

Spencer, Edgar Winston.
Geologic maps : a practical guide to the preparation and
interpretation of geologic maps / Edgar W. Spencer.—2nd ed.
p. cm.
Includes bibliographical references and index.
ISBN 0-13-011583-5
1. Geology—Maps. 2. Geological mapping. I. Title.
QE36.S64 2000
550′.022′3—dc21   99-33477
                                                          CIP

Senior Editor: Patrick Lynch
Executive Managing Editor: Kathleen Schiaparelli
Assistant Managing Editor: Lisa Kinne
Director of Marketing, ESM: John Tweeddale
Marketing Manager: Christine Henry
Manufacturing Manager: Trudy Pisciotti
Manufacturing Buyer: Michael Bell
Art Director: Jayne Conte
Cover Designer: Ann France
Cover Art: Davis Mesa Quadrangle, USGS, 1955, by F. W. Cater, Jr. (Courtesy USGS)
Production Supervision/Composition: Clarinda Publication Services

Printed in the United States of America
10 9 8 7 6 5 4 3 2 1

ISBN 0-13-011583-5

Prentice-Hall International (U.K.) Limited, *London*
Prentice-Hall of Australia Pty. Limited, *Sydney*
Prentice-Hall Canada Inc., *Toronto*
Prentice-Hall Hispanoamericana, S.A., *Mexico*
Prentice-Hall of India Private Limited, *New Delhi*
Prentice-Hall of Japan, Inc., *Tokyo*
Prentice-Hall (*Singapore*) Pte. Ltd.
Editoria Prentice-Hall do Brasil, Ltda., *Rio de Janeiro*

Contents

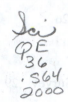

Preface

Geologic maps are among the basic tools used by anyone who wants to gain an understanding of the surface and shallow subsurface of the earth. They provide information about the types of materials that are present and the configuration of those materials in three dimensions. In addition, geologic maps identify the location of faults and rock structure. Some maps show the distribution of surficial materials, some depict only bedrock, and commonly, both are represented. In the hands of a skilled interpreter, geologic maps reveal the location of many types of natural hazards, indicate the suitability of the land surface for various uses, reveal problems that may be encountered in excavation, provide clues to the natural processes that have shaped an area, and lead to the potential location of important natural resources. For these reasons, civil and environmental engineers, land-use planners, soil scientists, and geographers, as well as geologists, use geologic maps.

This book is designed to provide instruction for students who are enrolled in map interpretation and field geology courses, but it may also be used for individual self-instruction by students or professionals who find that they need to understand and use geologic maps. To accomplish these goals, the book is written as a work manual. Following a brief discussion of basic information about map projections and the types of information generally presented on geologic maps, advice is given concerning the initial steps to be followed in map interpretation. The text covers maps showing surficial materials as well as bedrock geology. After the text describes representative examples of many of the types of features found on geologic maps, exercises direct the attention of students to those features in sections taken from published geologic maps. Geometric techniques are explained using a step-by-step approach.

Chapter 3 of the book provides basic information needed to prepare geologic maps. This chapter is designed for the student who is beginning a field mapping project. It will give those whose primary interest is map interpretation insight into the mapping process and an appreciation of the level of precision represented by data on geologic maps. Because aerial photographs are widely used in mapping, a short discussion of the use and interpretation of aerial photographs is included.

I gratefully acknowledge the help of many generations of students who have shared their experiences in learning to prepare and interpret geologic maps with me. Special thanks are extended to Ronald Erchul, Mary Westerback, Grenville Draper, and my daughter, Shannon Spencer, for their help in preparing the first edition; and to Marie Johnson, Edward Hansen, Peter Copeland, Jay Van Tassell, and Daniel Murray for their advice on the second edition. I greatly appreciate the help of Rena Thiagarajan, Christine Metzger, Andrea Creech, Andrew Thompson, Greg Bank, and Madelyn Miller for help in editing and preparing the manuscript for publication and of my daughter, Shawn Spencer, who prepared many of the new illustrations for this edition. Thanks also to Marcus Bursik of SUNY Buffalo, Edward C. Hansen of Hope College, Marie Johnson of United States Military Academy (West Point), Daniel P. Murray of the University of Rhode Island, Jay Van Tassell of Eastern Oregon University, and Peter Copeland of the University of Houston.

Edgar W. Spencer
Lexington, Virginia

About the Author

Edgar Winston Spencer is the Ruth Parmly Professor of Geology at Washington and Lee University where he has served as department head from 1959 until 1995. While a graduate student at Columbia University, he worked at the Lamont-Doherty Geological Observatory and taught at Hunter College. His dissertation concerned the structure of the Beartooth Mountains in Montana. He continued mapping and structural work in the Madison Mountains and later in the Appalachians where he has done regional mapping in the Blue Ridge and in the Valley and Ridge for the Virginia Division of Mineral Resources. He received an outstanding faculty award from the Virginia Council of Higher Education in 1991. He is a member of Sigma Xi and an honorary member of Phi Beta Kappa and Omicron Delta Kappa. Dr. Spencer is the author of a structural geology text and several introductory books. In recent years, he has continued research in the Central Appalachian Mountains, written guidebooks on the geology of that region, conducted field seminars for the American Association of Petroleum Geologists, and worked with alumni colleges.

CHAPTER 1

Maps and Images Used in the Study of Earth

Many techniques are used to portray the surface and near-surface features of the earth. Some of these, photographs and sketches of the landscape, depict the earth's surface in ways in which we are accustomed to viewing it. Other techniques, such as geologic maps and cross sections, are designed to reveal features that are not obvious to the casual observer. Each method of illustration has certain advantages. These maps or illustrations are the end product of the accumulation of large amounts of data, interpretation, revision, and documentation of the earth's surface. They allow the geologist, geographer, engineer, and planner to visually image this data and embark on an adventure to discern and understand the earth's surface and the underlying structure in an area of interest. This is the logical first step in understanding the natural environment and in deciding the need for additional geologic study or engineering work.

BASE MAPS

A base map is a map showing geographic and cultural features. Geologic data are recorded and presented on a base, most commonly a topographic map. The ideal base map is one drawn in such a way that the map contains a minimum distortion of horizontal distances and directions between all points on the ground. As you will see, the curved surface of the earth makes this a difficult task.

A number of different types of maps are used as bases for presentation of geological data. The most widely used bases in the United States are maps that depict topography by means of lines connecting points of equal elevation, called **contour lines.** Topographic contour maps are available at scales of 1:250,000; 1:100,000; 1:62,500; 1:50,000, and 1:24,000. Most of these are published by government surveys. Maps published by the United States Geological Survey are available from the Survey offices in Reston, Virginia, Denver, Colorado, and Menlo Park, California. Most recent maps published in other countries have scales of 1:25,000, 1:50,000, 1:100,000, or 1:250,000. The magazine *Geotimes* periodically publishes the addresses of state and national geological surveys throughout the world.

Base maps without contours are commonly used to depict large areas, e.g., states, regions, or an entire country. Bases without contours may also be used for maps containing data that might be confused by topographic contours.

If topographic maps are not available or if greater detail is needed than can be placed on the most detailed map available, aerial photographs taken with the camera pointing straight down, called vertical aerial photographs, may be used as base maps.

OBLIQUE AERIAL PHOTOGRAPHS

Photographs taken obliquely from the air (Figure 1–1a) retain some of the perspective of ground level photographs (Figure 1–1b). Most features are familiar and are easily recognized, but distortion, caused by change of scale with distance, remains.

FIGURE 1–1   (a) This oblique aerial photograph shows the collapsed crater of Mauna Loa volcano located on the big island of Hawaii. (Photograph from United States Geological Survey.) (b) Ground-level photograph of a massive cliff-forming sandstone layer underlain by thin-bedded sandstones and shales. Clearly identifiable lithologic units such as these constitute ideal rock units of the type shown on most detailed geologic maps. This photograph was taken in the Colorado Plateau region of the southwestern United States. Many rock units are not so clearly defined because the upper or lower contacts are gradational.

(a)

(b)

Because of this distortion, only vertical aerial photographs are suitable for use as base maps for geologic mapping.

VERTICAL AERIAL PHOTOGRAPHS

Many photographs used in the preparation of maps, and for photographic interpretation, are taken from high altitude and with the camera pointing vertically down (Figure 1–2a). These photographs have the advantage of showing features in their correct position relative to one another and with much less distortion than occurs in oblique photographs. Some distortion remains because the distance from the camera

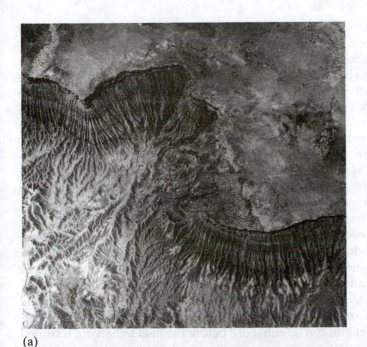

(a)

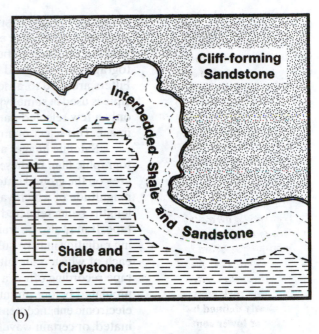

(b)

FIGURE 1–2   (a) This vertical aerial photograph is the type commonly used by geologists in mapping. The area shown is located in Utah along a prominent cliff (similar to that shown in Figure 1–1b) known as the Book Cliffs. (b) A geologic sketch map illustrates the aerial distribution of the three rock units shown on the vertical photograph. (Photograph from the United States Geological Survey.)

to the point on the ground shown in the center of the photograph is less than distances to points farther away from the center. Also, points on the ground at different elevations are distorted. These effects become less pronounced as the altitude of the camera increases.

ORTHOPHOTOGRAPHS

The limitations of vertical aerial photographs (i.e., radial distortion of scale from the center of the photograph to the edge) are corrected in orthophotos. The resulting photograph is planimetrically correct. Thus, accurate measurements of area, distance, and directions can be made on orthophotographs. For this reason, they make better base maps than other photographs, and they may be preferred in some instances over topographic maps for use as base maps. Many 1:24,000 quadrangles in the United States are now available as orthophotos. They may be obtained from the United States Geological Survey.

LANDSAT SATELLITE IMAGES

Images produced from remote sensing devices are used for mapping and monitoring the environment from satellites. These data are most helpful in revealing recent changes or new occurrences in an area. Images are especially important to environmental scientists and engineers who are studying time-dependent changes in the environment. Such changes may be natural or in response to a remediation technique. The devices used for this purpose are designed to detect radiation coming from the earth. This radiation can be characterized as a spectrum that ranges in wavelength from very long waves, such as radio, radar, and infrared waves, to short waves, such as ultraviolet, X rays, gamma rays, and cosmic waves. Some very short wavelengths can penetrate the earth's surface and provide information concerning buried objects and shallow structure. For example, images obtained from very short wavelengths can reveal the presence of buried stream channels beneath sand. Most films used in photography are designed to be sensitive to and record the visible part of this spectrum. Filters are used to absorb some wavelengths and emphasize others. This same principle is used in some types of remote sensing. Essentially, a photograph is taken using only a few selected parts of the radiant energy reaching the camera to produce the image. This selection may be made by use of special combinations of films and filters.

Satellite images are also obtained by use of a scanning system rather than photographic film. In these systems, a rotating mirror directs the radiation from a small area on the ground onto a detecting device, which generates an electrical impulse, the magnitude of which varies depending on the amount of energy of particular wavelengths being reflected onto it. The electrical impulse is then recorded on magnetic tape. It may also be transformed into a light beam and recorded on film. The incoming radiation may be subdivided according to wavelength into as many as 18 channels, each of which is simultaneously digitally recorded on magnetic tape. The magnetic tape recording offers a number of advantages, in that the signals from it can be manipulated before an image is produced. Such manipulations consist of filtering and electronic enhancement. In this way, various types of background "noise" can be eliminated, or certain wavelengths can be enhanced before the final signals are recorded as an image on a video system.

The scanning methods have been highly successful in providing relatively detailed images of large areas on the ground. They offer far greater flexibility than more conventional photographic methods and allow the user to enhance particular features by selecting and reproducing the radiation recorded in certain wavelengths during processing of the image. For example, selecting longer wavelengths of radiation (e.g., radar) results in excellent penetration of most clouds, haze, dust, and precipitation. Thermal infrared radiation (long wavelength) is emitted from warm and hot objects on the earth even at night, so images in this range obtained at night can be used to locate thermal springs, volcanic centers, and even other lower-level heat sources. Other wavelengths or combinations of wavelengths may be used to make air pollution, suspended sediment in water, various types of crops, or other surface features more prominent on the image.

Radiant energy outside the visible part of the spectrum is used in producing Landsat images. In the processing, each part of the spectrum used is arbitrarily assigned a color. The resulting image is made up of colors that are different from the ones people see. They are called false color images. For example, trees and green fields commonly appear red on these images. The advent of these techniques has made possible vast improvements in monitoring the environment and inventorying land-based resources.

Because the scale of most satellite images is so small (Landsat images with a scale of 1:1,000,000 cover areas of approximately 10,000 square miles—100 miles on each side), they are suitable mainly for reconnaissance mapping. Most geological maps continue to be prepared on more conventional base maps, such as topographic maps (1:24,000–1:100,000) and vertical aerial photographs.

SIDE-LOOKING AIRBORNE RADAR (SLAR) IMAGES

Using radar for purposes of detecting objects such as cars and airplanes is familiar to most people. In using radar, electromagnetic radiation with wavelengths commonly in the range of 0.5mm to 10m is directed outwardly. This radiation is thus much longer than the visible part of the spectrum. Usually the transmitter sends out a single wavelength. Part of that radiation is reflected from smooth objects or scattered from objects that have rough surfaces, and part of the reflected or scattered radiation is directed back toward the source and may be detected. To obtain images of the earth's surface, a radar source and detector are located in an airplane and directed toward the surface of the earth. The angle at which the detector is aimed can be varied to obtain energy returned at either a low or steep angle from the surface. The detector scans the surface, using a back-and-forth motion to detect returning radiation. These scan lines are recorded continuously as the airplane flies at a closely controlled altitude. The resulting image is a long strip oriented in the direction of the flight line. Strips can be placed together to produce a mosaic image of an area.

SLAR images resemble aerial photographs (Figure 1–3), but they are really quite different in a number of ways. The wavelength of radiation used for this purpose is much less affected by moisture and dust than is visible radiation. Thus, radar images contain no clouds. The longer wavelengths of radiation may even penetrate vegetation and dry sand or soil. Some images obtained in arid regions have successfully detected subsurface drainage systems now covered over with sand and dust. Like photographs, radar images cannot "see" the far side of objects. Thus, the back side of moun-

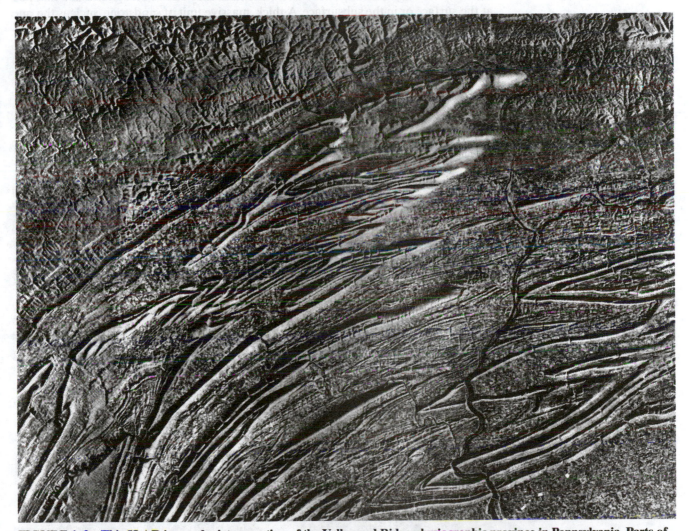

FIGURE 1–3   **This SLAR image depicts a portion of the Valley and Ridge physiographic province in Pennsylvania. Parts of the Great Valley (lower right), Valley and Ridge (central portion of the image), and Appalachian Plateau (upper left) are shown. Limestone forms the floor of the Great Valley. Folds developed in sandstone (ridges) and limestone and shale (valleys) form the dramatic topography of the Valley and Ridge Province. Flat-lying sandstones lie beneath the Appalachian Plateau. (Image compiled by Simulation Systems Inc. from data obtained from the United States Geological Survey.)**

tains or hills lie in shadows that are black on the images. The clarity of the images and the penetration of the radiation make SLAR images valuable sources of information.

GEOLOGIC MAPS

Geologists depict their interpretations of the aerial distribution of different rock bodies and surficial materials on maps called geologic maps (Figures 1–2b and 1–4c). The "bodies of rock" depicted may be bedrock materials such as sedimentary strata, igneous intrusions, or metamorphic rocks; or they may be surficial deposits, such as stream alluvium, beach deposits, or volcanic extrusions.

On some geologic maps, the bodies of rock that are identified and distinguished from one another are what geologists call **rock units.** These are bodies of rock that can be identified on the basis of their composition and texture. The basic rock unit is called a **formation.** These are bodies of rock that can be identified by their lithology and their stratigraphic position. By definition, they can be distinguished from the rock units stratigraphically above and below, and they can be recognized and mapped at the surface or in the subsurface. A thick, massive unit of sandstone, such as the one shown forming the cliff in Figure 1–2a, might be an example. Formations may be sub-

(a)

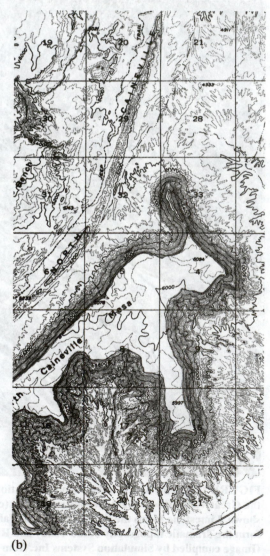

(b)

divided into thinner units called members; and in some cases, several formations that are related to one another are placed in larger stratigraphic subdivisions called groups.

On other geologic maps (e.g., the geologic map of the United States), the units differentiated on the map (called **map units**) are grouped on the basis of their age. For example, all sedimentary rocks of Cambrian age, regardless of composition, may be grouped together. Geologic maps always contain an explanation in which the units used on the map are identified and the symbols are explained. Common symbols used on geologic maps are included as an appendix in this book. Always examine the explanation to find out what is differentiated on the map.

The amount of control—that is, the number of places on the ground where observations were made—used to construct geologic maps varies greatly from map to map. In most areas, the number of places where rocks crop out at the surface limits the amount of control. The number of control points used in the construction of the map may also be determined by the amount of time available to collect data or the ease of access to outcrops. The contacts between different rock bodies appear as lines on geologic maps. In some areas, it may be possible to work out the position of contacts in great detail. In other areas, the contacts may be largely concealed from view, and their position may be inferred. Where bedrock is concealed, sources of information about the subsurface may be available from wells, borings, pits that have been dug, or from geo-

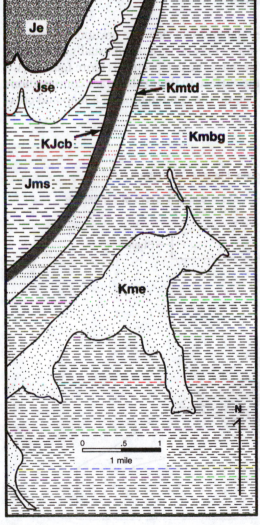

(c)

	Kme	Emery Sandstone
Cretaceous	Kmbg	Blue Gate Shale
	Kmtd	Tununk Shale
	KJcb	Cedar Mtn Fm
Jurassic	Jms	Morrison Fm
	Jse	Summerville Fm
	Je	Entrada Sandstone

FIGURE 1–4 (a) Vertical aerial photograph of an area in Utah covering part of North Caineville Mesa. (b) Topographic map of the area covered by the vertical aerial photograph shown in Figure 1–4a. (c) A geologic sketch map of the area shown in Figures 1–4a and b. North Caineville Mesa has a flat top and is surrounded by a cliff formed of the Emery Sandstone. The Blue Gate Shale surrounds it. The ridge labeled North Caineville is held up by a steeply inclined sandstone known as the Cedar Mountain Formation. Northwest of the ridge, rock units of Jurassic age are flat lying. (Photograph and topographic and geologic maps from the United States Geological Survey.)

physical surveys. Some geologic maps represent years of careful work on the ground; others are based largely on the interpretation of aerial photographs. Because geologic maps are interpretations based on a limited number of observations, locations of contacts and interpretations generally become more refined as an area is remapped and more detailed observations become available. Because geologic maps are drawn on a base map such as a topographic map (Figure 1–4b) or vertical aerial photograph (Figure 1–4a), it is possible to locate the geologic information in a geographic context.

GEOLOGIC CROSS SECTIONS

Ideally, a cross section (Figure 1–5a) shows the positions of contacts between strata or other rock bodies that you would be able to see if it were possible to make a vertical cut along a line across the ground. A profile of the ground surface is used for the top of a cross section, and the section may extend to any depth. If serious distortion is to be avoided, the same scales must be used for elevation (vertical) and horizontal distance. The position of features in cross sections is usually obtained by projecting what is seen on the ground beneath the surface. Well data, seismic data, and mine maps are also important sources of subsurface control, where they are available.

GEOLOGIC BLOCK DIAGRAMS

Block diagrams (Figure 1–5b) give a three-dimensional view of a portion of the earth. The perspective from which the block is drawn may be varied, but usually the surface and two sides of the block are depicted. Because the perspective is more familiar, we can see the relationships of features on the surface to their subsurface continuations more easily on block diagrams than we can on the combination of a map and cross

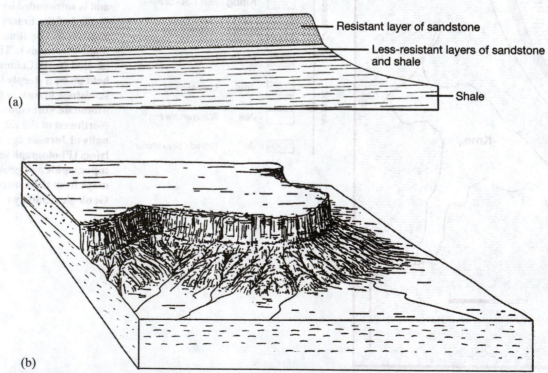

FIGURE 1–5   (a) A schematic geologic cross section of the area shown in Figure 1–5b. (b) This block diagram illustrates the area shown in the ground-level photograph of Figure 1–1b.

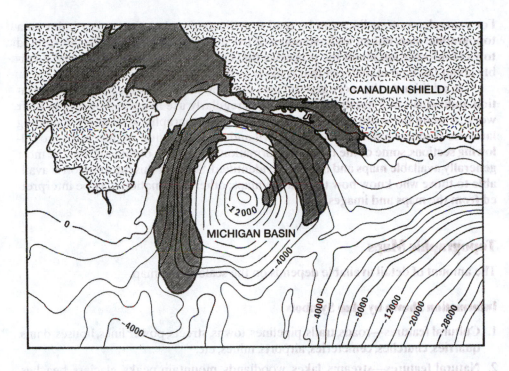

CANADIAN SHIELD

MICHIGAN BASIN

FIGURE 1–6   The Canadian Shield is a long-stable part of the North American craton where Precambrian igneous and metamorphic rocks are exposed. In the area south of the shield margin, Paleozoic sedimentary rocks cover the Precambrian crystalline rocks of the craton. The contours are lines connecting points of equal elevation on the top of the Precambrian crystalline rocks. The sedimentary rocks lie in a deep basin in Michigan. In the center of this basin, the Precambrian rocks are about 12,000 feet below sea level. Toward the eastern edge of this map area, the depth to the Precambrian crystalline rocks increases. Near the southeastern corner of the map, which is located in the Appalachian Basin, the Precambian rocks are nearly 10km (about 5.5 miles) below sea level. (After Alsoly, "The Basement Map of North America," AAPG ©1967; reprinted by permission of the American Association of Petroleum Geologists.)

sections. However, on blocks, both distances and angles are distorted. For accurate measurements, the combination of a map and cross sections is the best choice.

STRUCTURE CONTOUR MAPS

The shape of a rock or fault surface (commonly the top of a particular stratum) can be depicted by use of contours drawn to represent the surface (Figure 1–6). Such contours indicate the elevation relative to sea level of the surface of the rock body in the same way topographic contours show the elevation of the surface of the ground. Usually structure contours depict the elevations of rock surfaces beneath the ground.

TECTONIC MAPS

When the mapmaker's purpose is to emphasize the distribution of structural features rather than rock units, a tectonic map (Figure 1–6) is prepared. These maps typically show the locations of faults and fold axes. They may also show some important stratigraphic contacts, unconformities, igneous intrusions, or metamorphic terrain. Tectonic maps also commonly show structure contours.

TYPES OF INFORMATION YOU CAN OBTAIN FROM MAPS AND IMAGES

Maps and images contain a wealth of information. Much of this information can be obtained simply by reading the map—that is, by understanding the way the map is constructed and what the various symbols on the map represent. Much more information is available to those who have a more complete understanding of the subtle meaning of the patterns, shading, and configuration of contour lines and can interpret them in terms of natural processes and materials that are generally associated with them.

For example, contour lines can be read to indicate the elevation at any point on a topographic map, but an understanding of the shape of the land may be interpreted to yield information about the processes that caused the observed shape and possibly about the type of material that is likely to be found in certain landforms.

In addition to the more general uses of maps and images for purposes of location, individuals in a number of professions regularly use maps and images in their work. Among these are geologists; geophysicists; geographers; planners, including land-use planners and architects; and civil and environmental engineers. In the following sections, some of the types of information that can be obtained from the most generally available maps and images are identified. Some of this information is available to those who know how to read the maps; other information may be interpreted from the maps and images.

Topographic Maps

The amount of detail available depends on the scale of the map.

Information Shown by Map Symbols

1. Cultural features—roads, trails, pipelines, towns, streets, power lines, houses, dams, quarries, churches, cemeteries, airports, mines, etc.
2. Natural features—streams, lakes, woodlands, mountain peaks, glaciers, beaches, waterfalls, swamps
3. Political boundaries—national, state, county, city; and, in some parts of the country, townships, ranges, and section lines
4. Latitude and longitude of any point on the map
5. Scale showing horizontal distances
6. Elevation of the ground surface, indicated by contours and bench marks
7. Magnetic declination
8. Data of the map

Information You Can Interpret from the Map

1. Shape of the land surface (Profiles and block diagrams can be constructed.)
2. Types of landforms (A skilled interpreter can generally identify places where the landforms were created by erosion or deposition by glaciers, wind action, coastal currents, streams, and in some cases, by groundwater.)
3. Structure of the bedrock (For example, folds, faults, flat layers, etc. may be inferred from some maps.)
4. Drainage basins of streams

Geologic Maps

Information Shown by Map Symbols

1. Topographic information (If the geologic map is drawn on a topographic base, the information available on topographic maps of the same area is present on the geologic map, but contours may be difficult to read because colors are used to indicate geologic information.)
2. Type and location of bedrock units of various ages
3. Contacts between different rock units

4. Type and location of surficial deposits may be indicated
5. Type and location of faults and folds
6. Trend (strike) and inclination (dip) of rock layers

Information You Can Interpret from the Map

1. Rock structure beneath the ground surface, as indicated by cross sections
2. Rock type of the bedrock, both at the surface and at various depths in the sub-surface (This information can be projected from the surface.)
3. Rock hardness and consolidation (How difficult the rock will be to remove, if lithologies are known in detail.)
4. Origin and type of material in surficial deposits if the map shows surficial geology

LAND-USE MAPS DERIVED FROM GEOLOGIC MAPS

Many types of maps may be derived from the information provided on soil and geologic maps of bedrock and surficial deposits. An example is a map showing factors that affect land modification. Examples of the map units used on one such map (Warren County, Virginia, publication 15, Virginia Division of Mineral Resources, by Rader and Webb) follow:

Unconsolidated alluvium, human deposited fill
Unconsolidated pebbles, cobbles, and boulders in clay and sand
Shaly soil overlying interbedded shale and sandstone
Residual soil overlying shale and shaly limestone
Residual soil overlying limestone
Marked changes in soil thickness occurring over short distances
Acid, sandy soil underlain by sandstone and quartzite
Landslides
Karst areas with sinkholes

Based on the type of materials present and the slope in an area, it is possible to make certain inferences about the suitability of the land for various uses. For example, areas containing sinkholes generally have high potential for pollution of groundwater, and the surface of the ground is unstable. Scientists can also infer how susceptible the soil is to erosion, the stability of cuts and excavations, the probability and severity of movements of surface materials on slopes, the ease with which excavations may be made, the possible existence of important rock or mineral resources in the area, the limitations on development of the area for urban, industrial, or residential development, and the suitability of the land for waste disposal sites.

PROFESSIONAL USES OF GEOLOGIC MAPS

Geologists

1. To locate rocks of particular age, lithology, or structure.
2. To construct cross sections that will reveal the rock structure beneath the ground surface.
3. To reconstruct the geologic history of an area.
4. To explore for natural resources.
5. To locate water supply and groundwater recharge zones.

Civil and Environmental Engineers and Engineering Geologists

The basic information available concerns the type of materials that will be encountered at or beneath the ground surface and rock structure. This information is valuable for the following purposes:

1. Identification of natural hazards that may exist in any given area. This information is important in planning, design, and maintenance of engineering structures and in making environmental assessments.
2. To determine how difficult it will be to remove materials (e.g., can the surface be ripped, removed with earth-moving equipment, or will blasting be needed?).
3. To aid in evaluation of cost and problems that may be encountered in the site location for dams, building foundations, highway location, tunnels, canals, pipelines, and other structures.
4. In planning coastal structures and modification of or protection of shorelines.
5. Location of sites with bedrock suitable for waste disposal.
6. By understanding the type of earth material present in an area, an engineer can estimate its strength, deformation, and permeability characteristics. These physical properties must be verified by laboratory testing.

Planners and Architects

As they prepare plans for the future use of the land surface, planners need to have available much of the type of information that environmental engineers and geologists use in designing solutions to specific problems. The planning process provides an opportunity to avoid many of the site-specific problems that may arise as a result of failure to recognize potential environmental problems. Many of these problems are related to the character of the materials at or near the surface of the ground, surface and groundwater conditions, and the presence of natural hazards. Recognition of areas or features in the landscape that may influence the suitability of the land for various uses is important. For example, geologic maps enable identification of the following features:

1. Karst areas (sinkholes, caverns, disappearing rivers, etc.).
2. Flood-prone areas (if surficial geology is shown).
3. Areas where slope instability may exist.
4. Groundwater basins and recharge areas.
5. Active fault zones.
6. Geothermal areas.
7. Surface stream drainage basins.
8. The types of bedrock present and, for surficial geologic maps, the type of material that may lie on top of bedrock.

Soil Scientists

The composition and character of soil that has formed as a result of weathering of the underlying bedrock are closely related to that bedrock. Thus, geologic maps provide an important source of information concerning the origin and character of the soil. In the absence of detailed soil maps, geologic maps can be used to make generalized predictions about the character of the soil.

Base Maps

MAP PROJECTIONS

Maps are the most widely used method of depicting portions of the surface of the earth. While many maps are used to record the location of cultural features such as roads, buildings, towns, pipelines, or property boundaries, the types of maps used by geologists commonly depict the shape of the surface of the earth or the distribution of various rock types or rock units. Because of the problems involved in representing the surface of the earth—essentially a spherical surface—on a flat surface, maps generally contain certain distortions. Distortion of directions, areas, or both is inevitable, and the question facing the mapmaker is what type of distortion will present the fewest problems, considering the purpose for which the map is intended. If the area represented by the map is small, a few square kilometers or miles, the distortion may not be significant. But maps that cover large areas contain significant distortions. The amount of distortion generally increases with the size of area represented on the map.

The surface of the earth is subdivided by means of lines of latitude and longitude. Because the earth is nearly spherical, planes drawn through the equator or through the poles intersect the earth's surface in a circle. The center of the earth is the center of all such circles. Lines of longitude are imaginary lines located where planes that pass through the poles of rotation of the earth intersect the earth's surface. All lines of longitude are true north–south lines and are called meridian lines. The zero or prime meridian is the one that passes directly over Greenwich, England. Their angular distance (measured in degrees, minutes, and seconds) identifies other lines of longitude in the plane of the equator (Figure 2–1a), east or west from the prime meridian. From the sketch in Figure 2–1b, it is clear that lines of longitude converge toward the poles and are a maximum distance apart at the equator. Because topographic maps generally are bound on their east and west sides by lines of longitude, they are almost rectangular, but the width of the map is slightly different at the top and bottom.

Lines of latitude are lines formed by the intersection of the surface of the earth with planes that are parallel to the plane containing the equator. Unlike lines of longitude, these lines do not intersect one another. The angular distance of the line (measured in a meridian plane) distinguishes them from the equator. Thus, the latitude of a place on the surface of the earth is expressed as the number of degrees this place is north or south of the equator. Because all east–west lines on the earth's surface are parallel to one another, lines of latitude are commonly referred to as parallels of latitude.

Many different types of map projections are in use. Selection of a base depends on the size and location of the area to be depicted. Since no map projection is totally free of distortion, a choice is often made that will minimize distortion of area or direction or will keep the combination of distortions at a minimum.

FIGURE 2–1 (a) A cutaway view of the earth showing how latitude and longitude are measured. (b) An external view of the globe showing lines of latitude and longitude.

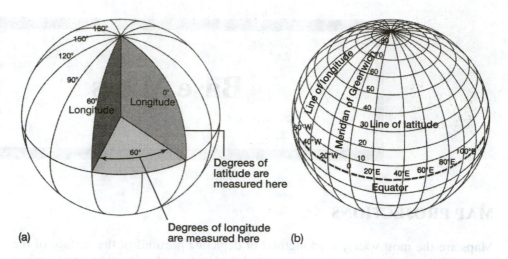

Mercator Projection

Most maps are rectangular or nearly rectangular in shape. One of the most widely used rectangular projections is one delineated by Gerardus Mercator in 1569, on which lines of latitude and longitude are laid out in a grid pattern. Lines of longitude (oriented north–south) are evenly spaced along the equator, and lines of latitude (east–west) are spaced farther and farther apart toward the poles (Figure 2–2a). Because of this distortion toward the poles, Mercator projections rarely show much of the polar regions. Despite this shortcoming, the Mercator projection is useful for navigation because bearings (compass directions) are straight lines on this projection.

The problems with the Mercator projection may be partially overcome by curving the lines of longitude toward the top and bottom of the map. This allows the areas to be kept under control, but at the cost of distorting directions.

Transverse Mercator Projection

This projection is similar to the Mercator, but the orientation of the cylinder on which the globe is projected is different (Figure 2–2b). Note that one meridian line on the globe touches the surface of the cylinder. Along that line and up to 15 degrees on either side, distortion is not excessive, but at greater distances from that line, distortion becomes a serious problem. This projection is used by the United States Geological Survey (USGS) for many quadrangle maps covering areas that range in size from 7.5 minutes to one degree.

Most maps published by the government contain tick marks that provide reference to a grid system known as the **Universal Transverse Mercator** (UTM) system. This grid system covers portions of the globe extending from 84° north to 80° south. This belt around the Earth is divided into 60 north–south zones, each of which is 6° longitude wide. The zones are numbered from west to east, starting at the 180° meridian. The point of origin for each zone lies along the equator at the point of intersection of a line of longitude and the equator. A separate grid exists for each of these zones. Points are located in terms of their distance east or west and north or south of this point of origin for the zone. On USGS topographic maps, tick marks, printed in blue, are spaced at 1,000-meter or 10,000-meter intervals along the map margins.

Polyconic Projection

Most base maps produced by the United States Geological Survey before 1950 used the polyconic projection shown in Figure 2–2c. Taking strips from the globe and flattening them out produces this projection, and stretching the outer part of each strip

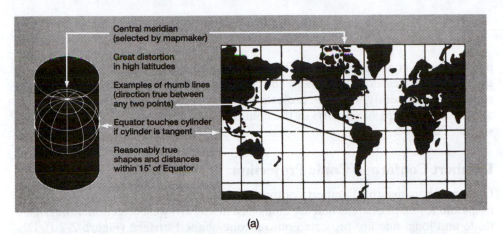

Central meridian (selected by mapmaker)

Great distortion in high latitudes

Examples of rhumb lines (direction true between any two points)

Equator touches cylinder if cylinder is tangent

Reasonably true shapes and distances within 15° of Equator

(a)

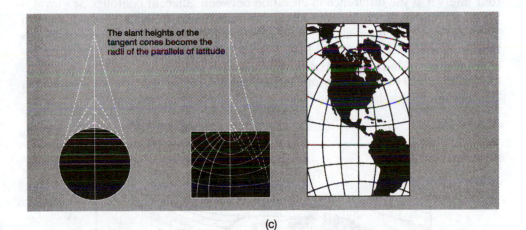

Central meridian selected by mapmaker touches cylinder if cylinder is tangent.

Equator

Can show whole Earth, but directions, distances, and areas are reasonably accurate only within 15° of the central meridian.

No rhumb lines

(b)

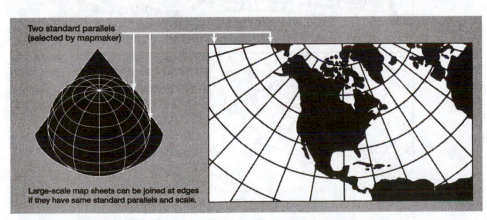

The slant heights of the tangent cones become the radii of the parallels of latitude

(c)

Two standard parallels (selected by mapmaker)

Large-scale map sheets can be joined at edges if they have same standard parallels and scale.

(d)

FIGURE 2–2 Four types of map projections: (a) the Mercator projection, (b) the transverse Mercator projection, (c) the polyconic projection, and (d) the Lambert conformal conic projection. All map projections contain distortion. Some distort areas; other distort directions. Note that areas of the same size on a globe vary in size as depicted on the Mercator projection. On a Mercator projection a surface area of any given size will appears six times larger at latitude 75 degrees than a comparable area near the equator. The transverse Mercator and the Lambert conformal conic projections are used for most maps of small areas published by the United States government. (From a poster published by the United States Geological Survey.)

until it forms a continuous surface. The scale along any line of latitude is constant, but the scale increases along meridians. The central part of this projection has little distortion. Consequently, when the projection is centered on the central United States, as shown, the maps of all parts of the country except Hawaii and Alaska are only slightly distorted.

Lambert Conformal Conic Projection

Cartographers use the Lambert conformal conic projection for many quadrangle maps and for maps of areas that are elongated in an east–west direction. Lines of latitude and longitude are projected onto a cone-shaped surface (Figure 2–2d). Distances are true only along two parallels of latitude, called standard parallels, where the surface of the cone intersects the surface of the globe. Distortion of directions and shapes is minimal. The standard parallels for the conterminous states in the United States are 33N and 45N.

TOPOGRAPHIC MAPS

The surface of the earth is represented on topographic maps by lines, called **contour lines,** that are map projections of lines connecting points of equal elevation on the ground (Figure 2–3). It may help to envision what a topographic map is like if you imagine that contours mark the position a lake shore would have if the land were slowly submerged, and the shoreline mapped when the water level reached successive elevations. The edge of a lake follows a contour line. To make the map easier to read, the contours are drawn at regular intervals. The interval, referred to as the contour interval, is the difference in elevation between adjacent contours. Sea level is the zero contour, and other contours are usually drawn at 5-, 10-, 20-, 40-, or even 100-foot (or meter) intervals. The contour interval is selected on the basis of the difference between the highest and lowest elevation in the area (which is known as relief), the scale of the map, and on the amount of elevation data available. Contours are commonly drawn at 5-foot intervals in areas where the relief is not great and at 100-foot intervals in high mountains.

FIGURE 2–3   Landscape drawing (top) and a topographic map of the same area. (From the United States Geological Survey.)

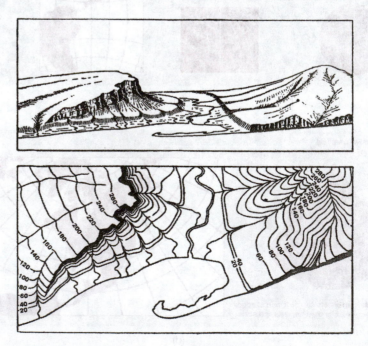

For small areas, contour maps are commonly prepared by conventional surveying techniques such as running a level line with a transit or level, but most topographic maps, such as those prepared by the U.S. Geological Survey or the Coast and Geodetic Survey, are made from vertical aerial photographs, supplemented by precise elevation control data obtained by surveying along roads and streams.

Location

Topographic maps produced by government agencies are generally named for a prominent locality (town or landmark). More precise information about the location of the map may be obtained by noting the longitude and latitude of the corners. Most of the recent maps produced in the United States cover an area that is 7.5 minutes of longitude wide by 7.5 minutes of latitude long. An older series of maps, which cover 15-minute areas at a scale of one inch to a mile (1:62,500), is also still widely available; and a third series covers areas two degrees (120 minutes) wide by one degree (60 minutes) long. Points within a map area may be specified in terms of their longitude and latitude, by means of the public land survey described subsequently, or in terms of their direction (bearing) and distance from a known locality.

Location by Means of the Public Land Survey

Most of the land in the United States has been subdivided by government surveys, as shown in Figure 2–4. The first step in these surveys was selection of an initial point (IP). This point was chosen at the intersection of a particular meridian and parallel, referred to as the Principal Meridian and Base Line, respectively.

The Principal Meridian and Base Line were then divided into 6-mile intervals, establishing a grid of townships, each 6 miles on a side. Each successive township north or south of the IP was given a township number, and each successive township east or west of the IP was given a number called the range number. Thus, Township 4 South, Range 3 East (T.4 S., R.3 E.) refers to the block between 18 and 24 miles south of the IP and between 12 and 18 miles east of the IP.

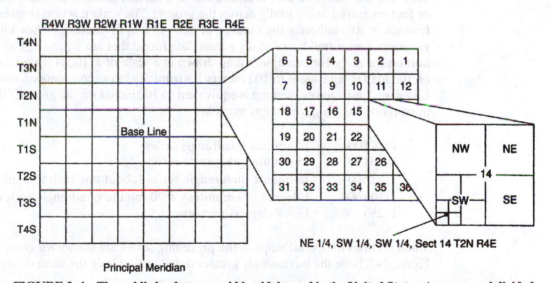

FIGURE 2–4   The public land survey grid is widely used in the United States. Areas are subdivided into townships and ranges. Each township is subdivided into 36 sections; each section is subdivided into quarters as shown. Because grids do not fit perfectly on the curved surface of the earth, many of the grid lines of this system are not perfectly north–south and east–west. For the same reason, sections are often not perfectly square. Note the section lines shown on some of the geologic maps in the appendix.

Since meridians converge northward (in the Northern Hemisphere), it was necessary to offset range lines periodically in order to maintain the 6-mile width of the townships. This was usually done at 24-mile intervals north and south of the IP along parallels called Standard Parallels. All north–south lines in the survey (range lines) except the Principal Meridian are offset in this manner. East–west lines (Township Lines, Standard Parallels, and Base Lines) are continuous.

Townships are subdivided into 36 sections or blocks of land measuring one mile on each side (one square mile in area). The system used to number sections within townships is shown in Figure 2–4. Location within a section is specified as closely as desired by quartering. By convention, the smallest quarter is given first. Thus, a point on a map might be specified as being in the NW quarter of the SE quarter of the SW quarter of section 10 in Township 9S Range 10E. This may be abbreviated as NW.SE.SW 10, T9S, R10E.

Ground Distance and Map Distance

We commonly measure distances by pacing, by using a tape measure that is placed on the ground, or by using a calibrated wheel that records distance as the wheel turns. Such distances are called ground distances. They are equal to the horizontal distance between points only if the ground is level. Thus, ground distances rarely correspond to map distances between the same points. The difference between ground and map (horizontal) distance increases with relief (Figure 2–5). Where precise measurements of horizontal distances are required, more sophisticated instruments than the ones just mentioned may be used. These include surveying instruments that measure the distance by using telescopes, the velocity of sound, laser beams, or other electronic methods. These instruments are generally used to measure horizontal rather than ground distances.

Map Scales

The scale of most quadrangle maps, expressed both as a graphic bar scale and a scale fraction, is given along the bottom margin of the map (Figure 2–6). Bar scales show the map distance that is equivalent to a certain number of miles, kilometers, or feet measured horizontally across the ground. The scale fraction (representative fraction = RF) indicates the number of units of length (meters, inches, kilometers, yards, etc.) measured horizontally across the ground that are equivalent to one such unit on the map. For example, a map drawn at a scale of 1:100 is scaled so 1 meter on the map is equivalent to 100 meters horizontally between points on the ground. Likewise, 1 inch across the map is equivalent to 100 inches on the ground. The scales in common use on government maps are

1:24,000	7.5-minute quadrangle series
1:50,000	15-minute quadrangle series
1:62,500	15-minute quadrangle series (about one inch to a mile)
1:100,000	1 degree (60-minutes) × 30-minute quadrangle series
1:250,000	1- × 2-degree map series

Small sections of several maps at the preceding scales are shown for comparison in Figure 2–7. Note the increasingly greater detail available as the scale changes.

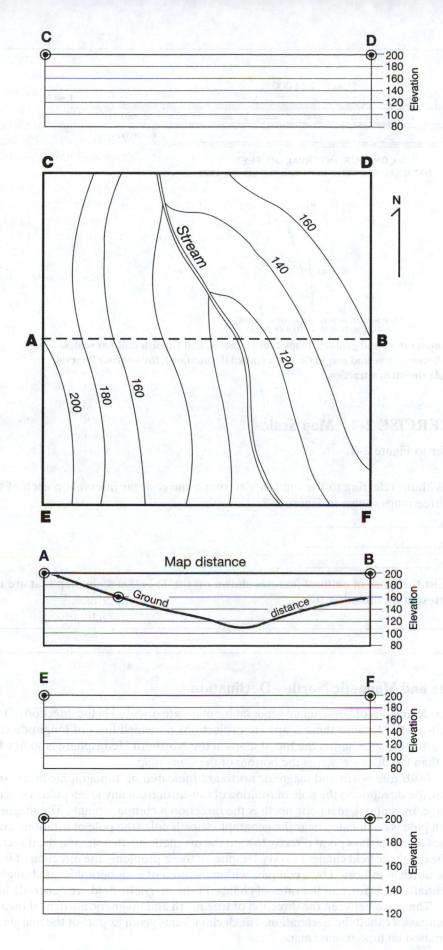

FIGURE 2–5 Center: A topographic map showing a stream valley and the lines along which topographic profiles are to be drawn. Bottom: A profile along the line A–B. Note that the distance from point A to point B is greater if measured on the ground than it is if measured on the map.

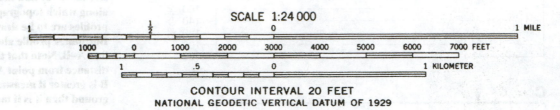

SCALE 1:24 000

CONTOUR INTERVAL 20 FEET
NATIONAL GEODETIC VERTICAL DATUM OF 1929

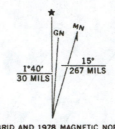

UTM GRID AND 1978 MAGNETIC NORTH
DECLINATION AT CENTER OF SHEET

FIGURE 2–6 Most topographic and geologic maps contain metric and English unit bar scales, the angular difference between true and magnetic north (called declination), the contour interval and datum, and the scale shown as a fraction.

EXERCISE 2–1 Map Scales

Refer to Figure 2–7.

1. Without referring to the caption, determine the contour interval on each of the three maps shown in Figure 2–7.

    a. _____

    b. _____

    c. _____

2. List the types of cultural features shown on the 1:24,000 scale map that are not present on the other two.

    _____

    _____

True and Magnetic North—Declination

The side margins of government topographic maps are oriented in the direction of true north–south. Because these maps are projections on which lines of longitude converge, the distance across the top of maps in the Northern Hemisphere is somewhat less than the distance across the bottom of the same map.

Both true north and magnetic north are indicated on topographic maps. True north, the direction to the pole of rotation of the earth from any given point, does not change. In contrast, magnetic north is the direction a compass points. The magnetic north pole is not located near the geographic north pole (the pole of rotation); compasses do not always point directly toward the magnetic north pole; and the direction of the magnetic field changes slowly. Despite all these problems, the direction of magnetic north as indicated by a compass within the area of a topographic quadrangle is essentially constant, and the rates of change in the magnetic field are generally low.

The angle between the direction of true north and magnetic north as shown by a compass is called the **declination.** The declination is given as part of the margin information on topographic maps.

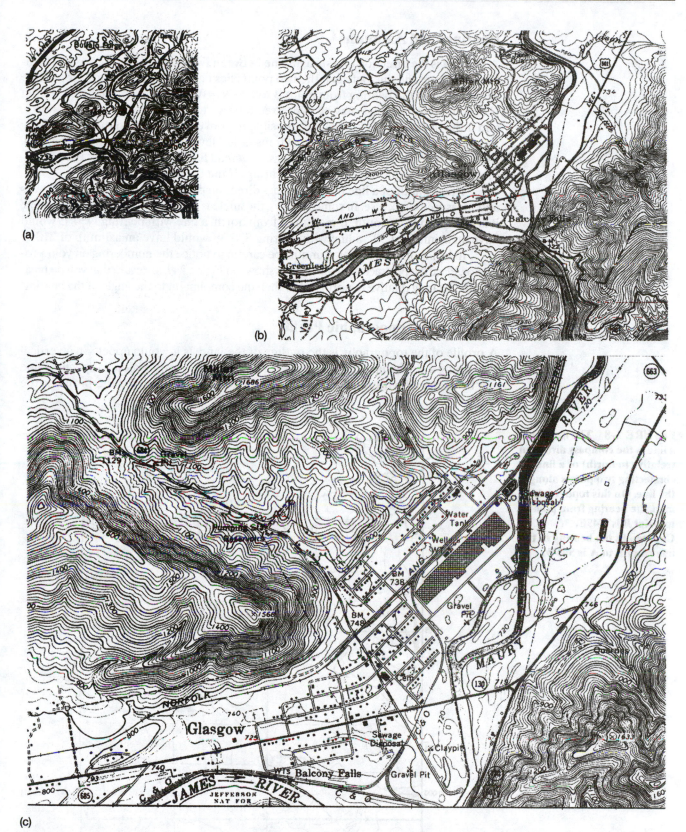

FIGURE 2–7   Topographic maps of portions of the same area. (Upper left) Part of a 1:250,000-scale map. (Upper right) Part of the area at left shown at a scale of 1:62,500. (Bottom) A small portion of the 1:250,000 map is depicted at the scale of 1:24,000. The contour interval is 20 feet on the 1:24,000 map, 50 feet on the 1:62,500 map, and 100 feet on the 1:250,000 map.

Bearings and Azimuths

A bearing is the direction of a straight line between two points, expressed as the number of degrees the line between the two points lies east or west of a north–south line (Figure 2–8). If the bearing from point B to C is N70W (Figure 2–8), the bearing from C to B is S70E. Bearings may be expressed relative to either true or magnetic north, and if the declination is known, it is possible to convert from one to the other. Many surveying instruments, including most transits and levels, measure magnetic bearings. These must be converted to true readings. The small hand-held compasses used by geologists can be adjusted to read true bearings if the declination is known.

The azimuth of a line is the compass direction of that line expressed in degrees and measured clockwise. In some surveys, the angle is measured from north; in others, it is measured from south. If measured from north, a line with bearing N30°E would have an azimuth of 30°; a line with bearing S30°W would have an azimuth of 210°.

In laying off bearings or azimuths, be careful to notice the numbering on your protractor. Place the protractor on the map, as shown in Figure 2–9, to ensure that you do measure the bearing relative to north rather than the compliment to the angle of the bearing.

Preparing a Topographic Profile

A profile of the topography is the outline of the land as it would appear in a vertical slice along a particular line (Figure 2–5). A topographic profile along a specified line can be prepared with relative ease if the horizontal scale used for the profile is the

FIGURE 2–8   The bearing of a line is the compass direction (relative to north) of a line connecting two points along the line. On this topographic map, the bearing from point A to point B is N45E. Conversely, the bearing of the line from B to A is S45W.

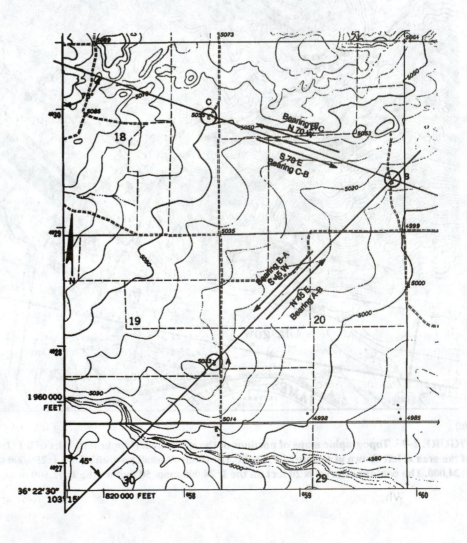

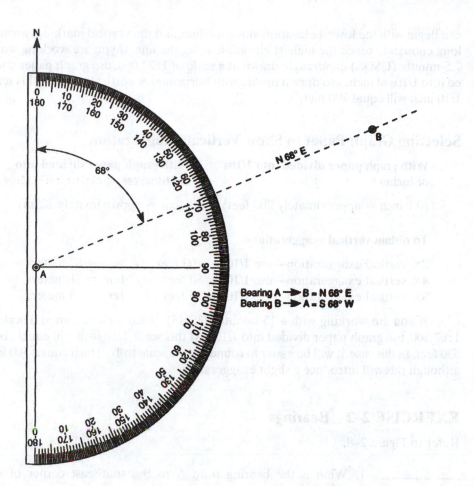

N

N 68° E

68°

A

B

Bearing A ➞ B = N 68° E
Bearing B ➞ A = S 68° W

FIGURE 2–9 A protractor is used to lay off a bearing from point A to point B. Be sure to set the protractor as shown, with 0 and the center point on the protractor oriented in a north–south line.

same as the horizontal scale of the map from which the profile is being prepared, and if graph paper with line spacing suitable for the scale is available. The most realistic representation of the topography is obtained when the vertical scale selected for the profile is the same as the horizontal scale. However, it is common practice to exaggerate the vertical scale, especially in areas of low relief, in order to make the features stand out more clearly. It will help you grasp the true dimensions of the land if a profile is prepared so the horizontal and vertical scales are the same, even if a second profile with exaggeration is to be prepared. Follow this procedure when constructing profiles.

Step 1. Select the vertical scale to be used and mark the scale divisions on a piece of graph paper. This example uses the same horizontal scale as that of the map.

Step 2. Place the edge of the graph paper along the line of profile on the map.

Step 3. Mark the point where each fifth contour (heavy contour) crosses the line of the profile, determine the elevation of each of these contours, and place a mark at that elevation. Also mark all stream crossings and ridge tops at the appropriate elevation. Finally, mark as many other contour crossings as are needed to clearly define the elevation of the land surface along the line of profile.

Step 4. Connect the points you have marked with a smooth line. Avoid using straight line segments unless you have reason to believe that the ground has a uniform slope. (The contours will be uniformly spaced for areas that can be represented by a plane surface.)

Note: When you are selecting a vertical scale, check the line of the profile and observe the highest and lowest elevations. The scale markings on your graph paper can begin with the lowest elevation along the line, and the vertical markings must be long enough to reach the highest elevation along the line. If you are working with a 7.5-minute (USGS) quadrangle drawn at a scale of 1:24,000, use graph paper divided into 1/10s of inches to draw a profile with horizontal = vertical scale. At this scale, 1/10 inch will equal 200 feet.

Selecting Graph Paper to Show Vertical Exaggeration

With graph paper divided into 1/10s of inches	With graph paper divided into centimeters
(0.1 inch = approximately 200 feet)	(1cm = approximately 125m)

To obtain vertical exaggerations:

2× vertical exaggeration—use 1/10″ = 100 feet	1cm = 62 meters
4× vertical exaggeration—use 1/10″ = 50 feet	1cm = 31 meters
8× vertical exaggeration—use 1/10″ = 25 feet	1cm = 15 meters

If you are working with a 15-minute (USGS) quadrangle drawn at a scale of 1:62,500, use graph paper divided into 1/10s. At this scale, 1/10 inch will equal about 530 feet. In this case, it will be easier to round off the scale to 1/10 inch equals 500 feet, although this will introduce a slight exaggeration.

EXERCISE 2–2   Bearings

Refer to Figure 2–8.

_____ 1. What is the bearing from A to the southeast corner of section 20?

_____ 2. Express the answer to question 1 as an azimuth (the number of degrees measured clockwise from north to the line of the bearing).

_____ 3. What is the bearing from the southeast corner of section 20 to the windmill at point C?

_____ 4. Express the answer to question 3 as an azimuth.

EXERCISE 2–3   Draw Topographic Profiles

1. Draw topographic profiles along the top and bottom edges of Figure 2–5.

2. On which of the three profiles is the ground distance across the map closest to the map distance? _____

3. Redraw the topographic profile along the line A–B using a 2× vertical exaggeration.

4. What are the advantages and disadvantages of using vertical exaggeration?

_____

_____

_____

Preparation
of Geologic Maps

Preparing a geologic map provides an excellent opportunity to combine skills in map reading, rock identification, measurement of structural features, and drafting. Before the map is complete, you will also use your abilities to visualize the three-dimensional shape of the rock units and how they relate to the form of the land. Because geologists prepare maps on the basis of a limited number of observations, the final product is an interpretation of what they can see and how those observations can be projected across covered areas. Finally, the map is a guide to what can be inferred about the subsurface and about the geological evolution of the area. This chapter outlines procedures and methods you may follow in carrying out a mapping project. A basic review of rock identification follows in Chapter 4.

PRELIMINARY PREPARATIONS

Define the Map Area

Before starting a mapping project, establish the boundaries of the area to be included on the map and draw these boundaries on a base map. It is always advisable to investigate immediately surrounding areas, and it is often helpful to map beyond the borders of the map area. You may find outcrops that will help you define the location of contacts, or you may discover faults or folds that may extend into your map area.

Collect and Review Existing Information

Locate as much information about the geology of the area you are mapping as possible. Look for earlier maps and descriptions of the rock bodies that are likely to crop out in your map area. This type of information may be obtained by talking with geologists who are familiar with the area, or by making a search of bibliographies of the geology of the region. The *Bibliography of Geology,* a reference work published by the American Geological Institute, is the most comprehensive list of references to published geological materials. World Wide Web sites operated by the federal or state geological surveys contain information about government documents and maps. The Expanded Academic Index and FirstSearch search engines contain excellent coverage of articles in periodicals and databases. Even if more detailed maps are not available, you should be able to find state geological maps that will provide general information about the structure and stratigraphic units that occur within the area you plan to map. If you are unable to locate information about the age and lithology of the rock units in the area, you will need to define these using the guidelines provided later in this chapter and in Chapter 4.

If you are undertaking a mapping exercise as part of a field course, you will either be given detailed descriptions of the rock units that occur within your map area, or you will be asked to define and describe the map units within the area.

Select a Base Map

Most geological maps of large areas are drawn on topographic map bases. Many geologists use both aerial photographs and topographic maps as base maps for location of places where data are being collected, even though the final map is prepared on a topographic map base. Because photographs record many features (e.g., trees, fences) that do not appear on topographic maps, it is frequently easier to determine your location in the field on an aerial photograph than it is on a topographic base. If photographs are used to record field observations, it is useful to use a pin to mark locations on the photograph. Then the location number or notes may be written on the back of the photograph and will not obscure the photographic image. When topographic maps are not available, aerial photographs are commonly used as a base. Orthophotographs have been corrected for distortion and make excellent base maps or maps on which to record data when contours are less important than recognition of landmarks on the ground surface.

The purpose for which the geologic map is being prepared influences the selection of a base. If the area being mapped is small and great detail is needed on the geology, a topographic map base may have to be prepared especially for the project. This is commonly the case where geologic maps are being prepared of mines or sites for building and dam foundations. Topographic maps may be prepared using surveying instruments, such as alidades, levels, or transits. The techniques are described in most surveying texts.

In the United States, most geological maps prepared by government surveys are prepared on topographic maps at scales of 1:24,000, 1:50,000, 1:62,500, 1:100,000, or 1:250,000.

MAKING A RECONNAISSANCE SURVEY OF THE AREA

A quick reconnaissance survey can save you considerable time in mapping. By making a quick tour of the area, you will have a good idea of where the best rock exposures are located, how to access various parts of the area, and where houses and property boundaries are located. You should make brief notes on a copy of your base map to indicate major changes in rock types and obvious structural features such as folds.

Obtain Permission to Enter Private Property

Be sure to introduce yourself to property owners in the area where you plan to map. Explain what you are doing and when you expect to be on their property. While many people are indifferent to trespassers, some consider trespassing a serious violation and may even have trespassers arrested. It is generally a good idea to secure permissions from as many of the property owners as possible before starting to map. Otherwise, you should be cautious about crossing poorly marked property boundaries.

COLLECTING AND RECORDING OBSERVATIONS

Decide Where to Collect Data

Deciding where to collect data will depend on how extensive the exposures of rock are in your area, access to those exposures, and the amount of time you have to complete the mapping. As a general principle, you cannot have too much data. But you may not have enough time to occupy every rock exposure in the area, and the lithology and structure of rocks may not change much from one outcrop to another. In such places, a few widely spaced observations may be sufficient. The scale of your map will determine how closely spaced data points can be placed on the map without having information overlap. **Your primary objectives are to identify and locate contacts between the rock units.**

Making Traverses.   Geologists generally use two techniques in deciding where to collect data. One technique involves collecting data encountered along selected traverses. The other technique involves finding and then tracing a contact across country wherever it goes. Generally geologists start mapping by making traverses. Initial traverses may be made along roads, along streams, or along ridges. It may be possible to complete the final map by correlating data from these traverses and scattered outcrops rather than by walking out each contact. However, if the structure of the area is complex, it may be necessary to trace out some contacts, but this can be a slow process. Generally, the final map is an interpretation of observations, some of which are located on contacts, but most of which are not.

You are likely to find outcrops in the following locations:

1. Along roads.   Select a starting point for the traverse that is easily located on your base. Knowing the scale of your base, you can locate points where observations are to be made by measuring with a tape, with a counting wheel, or by pacing. (You should determine your stride and pace the length by counting the number of steps you take to cover a previously measured distance. If you are going to use this method, make this determination before you start mapping.)

2. Along streams.   Pay careful attention to bends in the stream, places where tributaries shown on the base map enter the stream, rapids, sudden changes in gradient, or other natural features that appear on the base map (aerial photographs will be most useful where you are mapping away from cultural features). If you are uncertain about your location, check the bearing of straight sections of the stream and compare your reading with your base.

3. Along ridges.   Features that can be used for location on ridges may be scarce, but ridges afford good viewpoints from which distance features can be seen. Locate yourself by taking bearings on features you can identify on your base. Use the surveying method known as resection (see page 30) to locate yourself.

4. On slopes.   Be especially careful about location of points when you are traversing up or down a mountain slope. It is easy to overestimate the elevation you have gained in climbing upslope. Use an altimeter as a way of checking your location on a slope. But remember that altimeters are sensitive to atmospheric pressure changes, so they may not be reliable if the weather is changing rapidly. As a precaution, set your altimeter at a place to which you will return when your traverse is completed. If the altimeter does not give the same reading at this place at the end of the traverse, a change in atmospheric pressure has taken place. You will also know if the pressure increased or decreased. It may be even more helpful to check the altimeter whenever you are at a known elevation. Reset the altimeter where elevation is known and make corrections for elevations if changes have occurred since it was last reset.

Make corrections for altimeter readings as follows: (a) Set altimeter and record time while you are located at a point of known elevation (e.g., a bench mark or BM). (b) Read altimeter and record time at each observation point. (c) Record the amount by which the altimeter is off when elevation is again known, and record time. (d) Assume that the error has accumulated at a uniform rate. (e) Plot change in elevation against time. (f) Correct elevations at observation points using the correction indicated on the plot for the time at which the observations were made.

Tracing Contacts.   Following the contact between rock units, or zigzagging across a contact is the best possible way of determining the trace of a contact. Unfortunately, this method is also time consuming, and in many areas outcrops are so scattered that it is not a practical way of mapping. In some areas, contacts may be clearly identifiable on aerial photographs or other types of images. In such areas, the position of the contact should be verified in the field and transferred from the photographic image to the base.

Record Observations

Record your observations in a field notebook. Notebooks containing water-resistant paper are available through many outlets. These are obviously valuable if you are working in an area where rain is likely. Be sure to use a pencil or waterproof ink.

The base map you are using in the field should be protected from damage by rain or other hazards. It is convenient to mount the map (in sections if it is large) on hardboard, Plexiglas, or some other firm backing. Aluminum cases are available commercially. If the map is small or can be cut into small sections, it may be more convenient to mount these small pieces in your notebook.

Key the notes you take to a point on the base map. Place a point (it is often useful to place a small circle around the point) on the base, and record the key number beside it. Be certain that the number you select has not been previously used. To avoid this problem, it is advisable to number all observations within a single map area consecutively. Do not start a new set of numbers for each day, week, or month. To do so is likely to lead to chaos in the end!

Keeping a Field Notebook.   In your notebook, record the key number, the date, and describe the location of the point as precisely as possible. In making geological observations, assume that you will never return to that location. Record all of the information you may later need. That will include a complete description of the rock types present, the strike and dip of bedding, and other structural features (e.g., cleavage) that are present. If you know the name of the rock unit present at this location, record that. If you are uncertain about the identification, be sure to record that. Be sure to note any evidence indicating whether the strata are right side up or upside down.

Determine Your Location

It is essential that you know where you are located when you make and record field observations. Incorrectly located data will almost certainly complicate your interpretation and may cause you to misplace contacts, or even infer faults or folds, that are not present in the area.

The degree of precision needed for a map project will determine which methods are suitable for locating data points. The degree of precision needed depends on the purpose for which the map is being prepared and the scale of the map. For example, a geologic map of a quarry or a building site covering a few thousand meters will likely require a much higher precision than a map covering hundreds or thousands of kilometers of countryside. High-precision maps generally require the use of plane table and alidade, or other surveying techniques that are beyond the scope of this book. See the reference to surveying books.

Using Features on Base Maps or Aerial Photographs.   This is the simplest method of locating points where you plan to collect data. In using this method, be careful to determine the year in which the base map or photograph was made. Cultural features may change dramatically over short periods of time as people build new houses, driveways, or fence lines, and as highway departments change the positions of roads and bridges. Even natural features may change as forest lands are cut and stream channels modified by engineering works. If you are using aerial photographs, be sure to note the time of year when the photograph was made.

Measuring Distances.   Unless you are at a point of intersection of two features (e.g., roads or streams), you will need to determine how far you are from a feature you have positively identified on your map. You can obtain a rough estimate of that distance by pacing. To do this you must first know the length of your stride or pace. Using a tape measure, lay out a line of known length (30 meters or more). Walking with a comfortable stride, count the number of strides you take to cover that distance, and then calculate your average stride length. A pace is two stride lengths. For long distances you may wish to use a pedometer—an instrument that counts the number of paces. For a more accurate measurement of ground distance, use a tape or a distance measuring wheel. For even higher accuracy, use an optical or laser rangefinder. A number of models are commercially available (see the list in the reference section). A laser rangefinder may be accurate to within one meter at a distance of several hundred meters. Note that the distances measured with optical or laser rangefinders are horizontal distances. Distances determined by pacing or placing a tape along the ground are ground distances—not map distances.

Pace-and-Compass Surveying.   For field work that does not require a high degree of precision, geologists commonly use a compass to determine the bearing of lines, and they pace distances between points on the survey. You can get a good impression of the accuracy of a pace-and-compass survey by surveying a closed loop. After the survey is made, points are plotted on a base map. If the measurements are accurate the last point will coincide with the initial point.

When the area to be mapped is so small that topographic maps or aerial photographs are not suitable for use as base maps, a new base map must be prepared. When the level of precision required is not great, you can prepare a map by using the pace-and-compass method. In the pace-and-compass method, the Brunton compass is used to determine bearings, and distances are measured by pacing or by measuring with a tape. A method known as resection may also be used to determine the location of points that are not readily accessible or where distances cannot be accurately measured by use of pacing or taping (e.g., where the point to be located is not on level ground).

EXERCISE 3–1   A Pace-and-Compass Survey

Start by determining your pace (you may want to practice along the edge of a football field). Once you are satisfied you know your pace, make the following survey.

1. Map the corners of the football field.
2. Locate additional points close to the field.
3. Draw a map of the field and show any additional points you located. Use a piece of graph paper using a scale of 0.5 inch = 100 feet. Orient the map relative to north, and draw a "declination" diagram and scale. Express the scale as a fraction.

Using Global Positioning Systems.   By utilizing signals received from satellites, these instruments can determine your latitude and longitude with an accuracy of 3 meters or less. GPS is now widely used for location determination, but the cost of instruments capable of producing highly accurate results may limit their availability to some field workers.

Location by Resection.   If you are located at a point from which two or more known points (points that you can recognize on your base map) are visible, you can locate yourself by using a surveying technique known as resection. To locate yourself, take a bearing on the two known points. Then draw a line on your base map using the reverse bearings. (If the bearing from your location to a known point is N30E, the reverse bearing from that point to you is S30W.) You are located at the point where the two reverse bearings intersect.

Caution. A small error in measuring the bearings to these two points can result in a large error in your location. The accuracy of resection increases as the angle between the two lines used to make the determination increases.

DESCRIBING THE OUTCROP

When you leave an outcrop, your field notebook and base map should contain a precise location of the outcrop, a description of the type of rock or rocks present at that location, and information about the structure of those rocks. See the separate section devoted to the description of rock types in Chapter 4. Description of the rock structure should always include the strike and dip of sedimentary layers, the orientation of the axis and axial plane of folds (see Chapter 10), and any evidence you see of faulting (see Chapter 11). Your final geologic map will show the contacts between igneous, metamorphic, and sedimentary rocks. Each of these types of rock may be further subdivided into what geologists call map units or rock units.

Stratigraphic Units Used on Geologic Maps of Bedrock

Sedimentary rocks compose the bedrock over nearly three-quarters of the land surface of the earth. Most of these were deposited as sediment on the sea floor. In general, marine sediments form through the settling of matter through water to the sea floor, where a layer is built up. If the material being supplied or the physical, chemical, or biological factors influencing the environment change, the nature of the layers being formed also changes. Similar processes operate in lakes and other freshwater bodies. These processes of sedimentation bring into existence a feature known as stratification, a structure produced by deposition of sediments in beds, layers, laminae, lenses, wedges, and other essentially tabular units. The sediments may be fragmental or crystalline; formed from the settling of solid, insoluble debris; or the product of chemical precipitation. They may be composed of organic remains, or may be the result of some combination of these processes. Stratification results from variations in color, texture (size, shape, and fabric), density, composition, or fossil content from one layer to the next. If the layered rock can be recognized in the field and mapped as units on the basis of objective criteria, it may be used to define a **rock unit** (Figure 3–1).

Recognition of rock units is so important that geologists use a standard of nomenclature to describe them. Here are the various types of rock units:

> Groups
> > Formations
> > > Members
> > > > Lenses—lentils—tongues—beds

The basic unit, the **formation,** is a lithologically distinctive product of essentially continuous sedimentation selected from a local succession of strata as a convenient unit for purposes of mapping, description, and reference. In general, one or two constituents, such as massive layers of sandstone or thick units of alternating

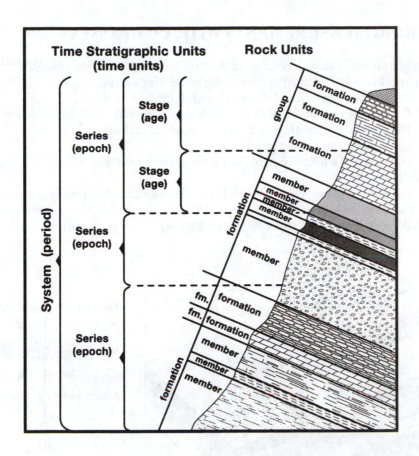

FIGURE 3–1 In defining rock units for purposes of mapping and study of stratigraphy, geologists subdivided stratigraphic sections into formations, members of formations, and groups of formations. Not all formations are subdivided into members; nor are all formations parts of groups. Geologic time is subdivided into eras, periods, epochs, and ages, and the rocks formed during each of these time subdivisions are called systems, series, and stages. The rocks formed during time subdivisions are time stratigraphic units.

shale and sandstone, dominate the composition of a formation. Formations usually have two names; the first is the name of the locality where the unit was first described or the name of a place where it is exceptionally well exposed, called the type section. The second name, when it is used, gives the rock type, if the formation is composed mainly of a single rock type (i.e., Martinsburg shale, Tuscarora sandstone, or Antietam quartzite). If the lithology is not so distinctive that a single rock type name can be used, the geographic name is followed by the word *formation*.

Groups.   If a number of formations in a sequence have similar lithology or were formed in a closely related environment, geologists may refer to the sequence as a group (i.e., Clinton Group, Medina Group, Chilhowee Group). In some instances, unconformities separate a group from over- and underlying formations.

Members, Beds, Lenses.   In stratigraphic studies and where mapping is done in detail, it may be desirable to subdivide formations. These subdivisions are called members if they extend over a large area; they are called lentils or lenses if they are only locally distributed; and they are called tongues or wedges if they wedge out in one direction between sediments of different lithology. The terms *beds* and *laminae* are also used for these smallest subdivisions.

Although all geologic maps illustrate the geographic distribution of natural sediment and rock materials at or near the ground surface, most maps show formations. More detailed maps may show members, beds, or lenses. The type of subdivisions used depends on the stratigraphy of the area, as well as on the scale and the intended use of the map.

MAKING MEASUREMENTS WITH COMPASSES

Geologists use compasses to measure the orientation of lines and planes. The orientation of a line is determined by measuring its bearing and plunge. The orientation of planes is described in terms of their strike and dip (Figure 3–2).

The **bearing or azimuth of a line** is the compass direction of that line (Chapter 2).

The **plunge of a line** (Figure 3–3) is the angle of inclination of the line measured in a vertical plane containing that line.

The **strike of a plane** (Figure 3–2) is the compass direction (bearing) of a horizontal line on the plane.

The **dip of a plane** (Figure 3–2) is the angle of inclination of the plane, measured in a plane that is perpendicular to the strike. The direction of dip must always be indicated. *Note:* Planes that strike northeast may dip either southeast or northwest, so

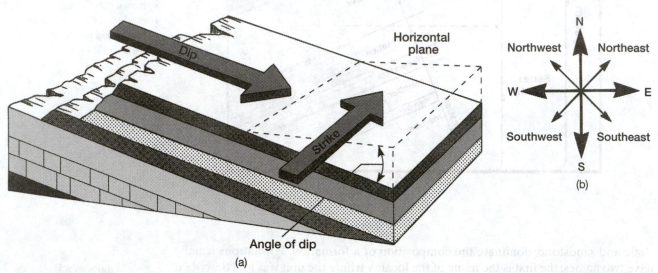

(a)

(b)

FIGURE 3–2   (a) The strike and dip of a bed are indicated on this block diagram. The direction of dip is at right angles to the strike direction. (b) Sketch showing the quadrants of the compass. This can be used to help visualize the possible directions of dip if the strike is known.

FIGURE 3–3   Define a horizontal line by measuring its bearing (see Chapter 1). Define an inclined line by measuring the bearing of its horizontal projection and its plunge, the angle of inclination of the line measured in a vertical plane.

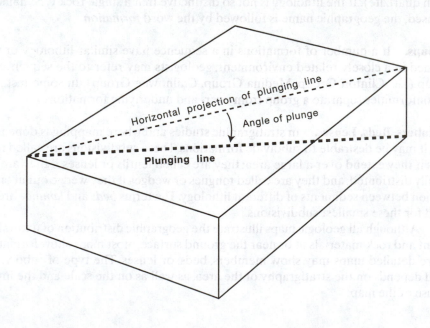

the direction must be indicated (Figure 3–2). Similarly, planes that strike northwest may dip either northeast or southwest.

Several different types of compasses are available, but two types of compasses, the Brunton and Silva (Figures 3–4 and 3–5), are most widely used by geologists. Both of these are designed to make measurement of compass direction and inclinations simple, but they are slightly different from most other compasses in that they contain level bubbles, a mechanism for measuring angles in a vertical plane, and in that east and west are reversed (Figure 3–4).

Because the orientation of lines and planes is generally plotted on base maps relative to true north, it is desirable to measure true bearing. Both of the compasses allow you to do this.

Declination is the angle between magnetic and true north. Declination varies by more than 20 degrees in the United States. Consequently, all measurements shown on geologic maps are shown relative to true north rather than magnetic north. The needle in a compass is aligned relative to magnetic north; therefore, you must either take magnetic readings and convert them to true readings before plotting them, or you must have a compass that can be adjusted to permit readings relative to true north. It is possible to make this type of setting on both Brunton and Silva compasses (Figures 3–4 and 3–5). Setting declination should be among the first steps you take before starting field work in an area. Check and, if necessary, reset the declination on the compass. Small screws are used (see drawings) to make these adjustments.

To set the declination on a compass, find out the declination in an area by obtaining a topographic map of the area. Declination is shown along the bottom margin. To set declination correctly, (a) set the declination to zero, (b) take a bearing on a point, (c) correct the magnetic bearing reading by adding or subtracting the declination from it (in this way, you will know the true bearing), and (d) set the declination on the compass so that bearing to the same point gives a true reading.

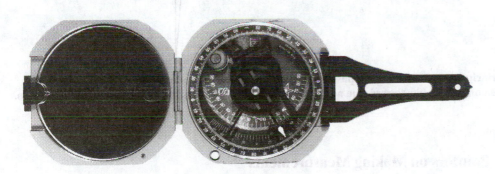

FIGURE 3–4 Photograph and sketch showing the principal parts of a Brunton compass.

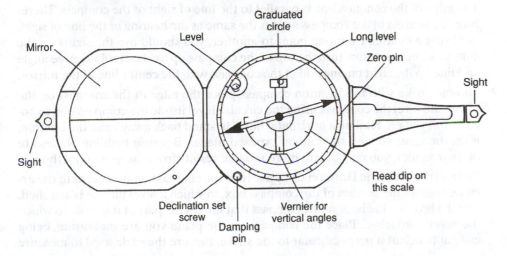

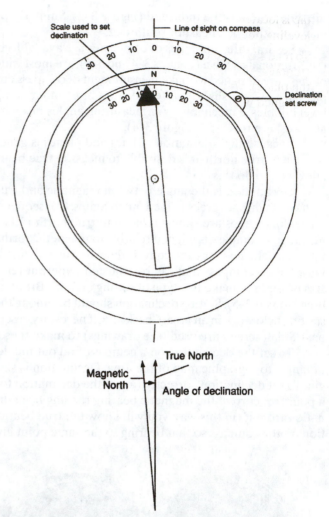

FIGURE 3–5   Sketches of the principal parts of a Silva compass. Adjusting the set screw sets declination. The set screw is turned until the line of sight of the compass points toward true north when the north end of the magnetic needle points to zero.

Pointers on Making Measurements

1. The edge of the compass box is parallel to the line of sight of the compass. Therefore, the bearing of the compass edge is the same as the bearing of the line of sight. In taking a bearing from one point to another, you should use the sights on the compass. On a Brunton, turn the top of the compass up to about a 45-degree angle and line up the sight on one end of the compass with the center line of the mirror.

2. To read strike with the Brunton compass, place the edge of the compass on the plane and level the compass, using the circular level inside the compass box to ensure that you are reading a horizontal line. Or, stand back away from the outcrop, level the compass, and sight along exposed planes. Because bedding planes are often irregular, you may obtain more reliable results from the second method.

3. To read dips with the Brunton compass, turn the compass on its side; note the arcuate scale in the bottom of the compass box, to which a level bubble is attached, and the lever on the back of the compass that moves the part of this scale to which the level is attached. Place the compass on the plane you are measuring, being careful to orient it perpendicular to the strike. Be sure the scale used to measure

dip is located in the bottom half of the compass. Then rotate the scale until the small leveling device is level, and read the dip on the fixed scale.

4. To read strike with the Silva compass, place the compass on the plane as nearly horizontal as possible (or align the line of sight with the trend of a horizontal line on the plane). Next, rotate the outer circle until the arrow in the bottom of the box is aligned with the compass needle. Then read the bearing of the line from the circular scale around the edge of the compass.

5. To read dips with the Silva compass, first turn the outer circle until the E–W markings on the outer circle are aligned along the line of sight of the compass. Place the compass on the plane you are measuring and align it in a plane that is perpendicular to the plane you are measuring. The small red pointer will swing into a vertical position. Read the dip on the arcuate scale in the bottom of the compass box.

Common Problems in Measuring Strike and Dip

1. Identifying bedding.   Bedding planes are usually easy to see in sedimentary rock sequences in which layers have different colors or are of greatly different composition. However, difficulties may arise when the changes in composition are slight; when units are thick, making massive beds hard to see; or when bedding may be confused with rock cleavage or even fractures. Remember that strata are distinguished from one another by changes in color, texture, or composition. These changes may be quite subtle. Keep looking until you feel confident of what you are measuring. Faulty readings will make interpretation of the data confusing or will lead to errors in interpretation. It is better to have no reading than to have one that is incorrect!

2. Finding a horizontal line on the bedding plane.   The Brunton compass contains a level bubble in a circular well. When this bubble indicates that the compass box is level, the edges of the compass are horizontal lines. Place one of the side edges on the plane you are measuring, and you should have no problem. If you continue to experience this problem, try placing a pencil on the plane in a horizontal line, or get a small carpenter's level and use it to find a horizontal line on the plane.

3. Indicating the correct dip direction.   Remember that the plane cannot dip in the same direction that the plane strikes. If, in recording your data, you see this error, look again to determine the direction of dip. It may help to stand away from the outcrop and look in the direction of the strike. If the plane dips to your right while you face northeast, the plane must dip southeast; if you are facing northwest and the plane dips to your right, it must dip northeast.

4. Sighting along strike.   When you are standing in line with the strike of a bed and looking in the direction of strike, you will be unable to see the top or bottom surface of that bed. If you are off to the side of the strike direction, you will be able to see portions of the top or bottom of the bed. Move until you are in the line of strike. Then measure the direction from your position to a point on the bed at about the same elevation as the compass.

COMPILING FIELD OBSERVATIONS ON THE BASE MAP

Field data initially recorded as key points on field maps should be transferred to base maps that are the same scale and type as the final product. A good technique involves covering a copy of the final base map with a sheet of mylar or other suitable transparent plastic material on which data can be recorded (Figure 3–6). Mylar is durable, and it is easy to erase. Professional geologists commonly obtain mylar copies of the topographic base maps. Mylar is a stable base and will not change dimensions as paper does when humidity and temperature change.

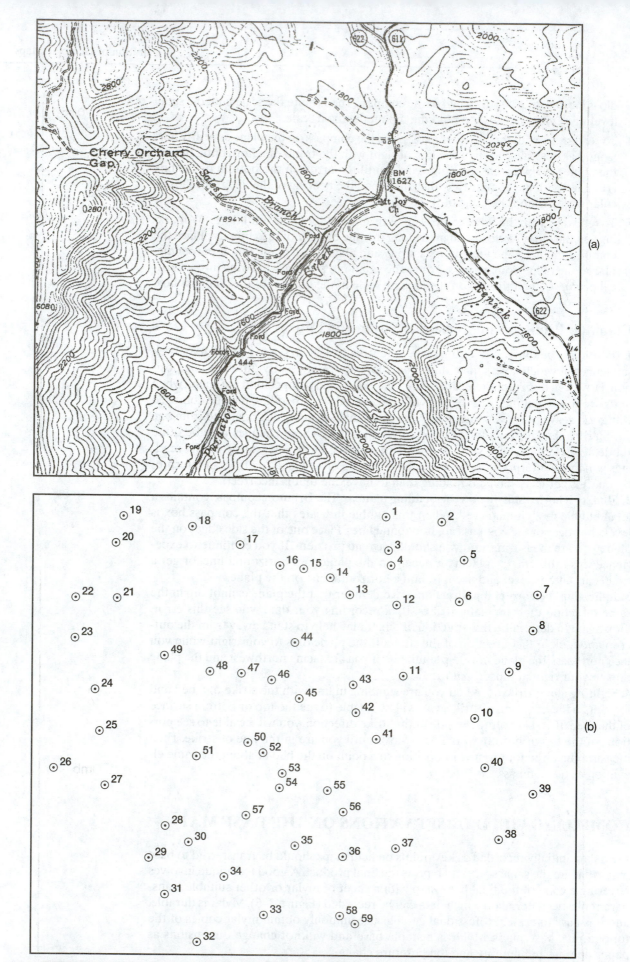

FIGURE 3–6   Steps in the preparation of a geologic map. (a) A topographic map is generally used as a base. (b) Localities at which data are collected are marked on the topographic map or on an overlay. For each of these localities, the type of rock, strike and dip data, and other information is recorded in the field notebook.

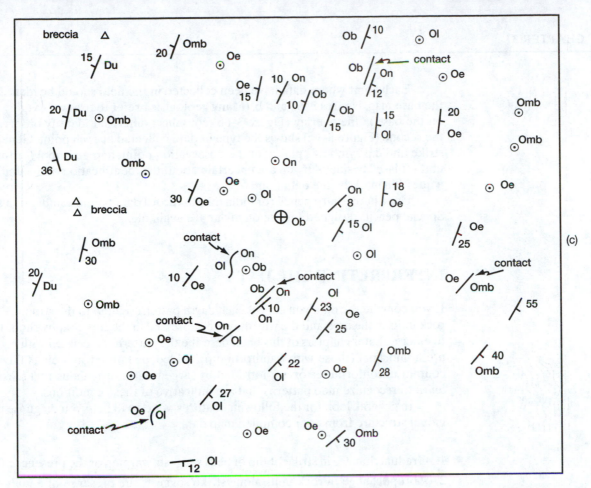

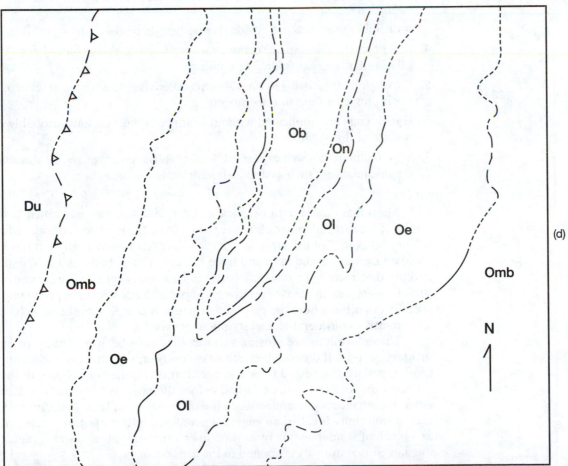

FIGURE 3–6 **(c) Data from the field notebook are plotted on the base map for each locality. In some places, strike and dip information is available. (d) Contacts based on the data shown in (c) are sketched. The final map should show representative strike and dip data as well as contacts, a north arrow, and a scale.**

Each point where data have been collected in the field should be identified on the base map, Figures 3–6a and b. (Many geologists prefer to use two overlay sheets on the base.) One overlay (Figure 3–6b) contains points that identify observations; the second (Figure 3–6c) shows the type of data collected at each point. Generally, a strike and dip symbol is placed at the observation point, and a symbol for the rock unit is placed beside it. If you are uncertain about the identification of the unit, place a question mark by the data point.

Use a hard, sharp pencil (or india ink) to record data on the compilation sheets. Special pencils designed for use on mylar are available.

INTERPRETING THE DATA

If you compile the data you collect each day, a pattern related to the structure of the rock units in the area and the way they are expressed in the topography should begin to emerge. Later chapters of this book describe the patterns associated with beds that are homoclinal (those with a uniform dip), folded, or faulted, as well as those that contain unconformities or are intruded by large bodies of igneous rocks. You will learn to recognize map patterns that are indicative of these conditions.

In general, look for the following features, which will help you diagnose the regional structure from your compiled map data:

1. Gradual changes in strike or dip of beds suggests warping or the presence of folds.
2. Abrupt change in rock units along strike suggests the presence of a fault or unconformity.
3. Indications that beds are upside down suggests faults or strong folding.
4. In a normal stratigraphic section, younger beds lie above older beds. Any reversal of order suggests folding or faulting.
5. If rock units are missing in what appears otherwise to be a normal section, you have likely crossed a fault or unconformity.
6. If rock units are duplicated without changes in dip, you have probably crossed a reverse or thrust fault.
7. Most faults are not well exposed. Usually, faults are recognized by abnormal stratigraphic relations, such as duplication or omission of section.

Maps that depict both bedrock and surficial materials are much easier to interpret if the surficial materials are removed from the map and considered separately. This "stripping" of the surficial materials may be done mentally, but it is helpful to place an overlay on the map and trace the contacts of bedrock units, faults, etc. In making this tracing, the contacts between beds and faults are projected under the surficial materials. In areas of relatively simple bedrock structure, projecting contacts is easily done, but where the geologic structure is complex or the surficial materials are extensive, considerable uncertainty may exist.

Unconformities are erosion surfaces that have been buried by younger sedimentary deposits. If the rock beneath an unconformity is igneous or metamorphic, the erosion surface is referred to as a nonconformity. If the rocks beneath the erosion surface were folded, tilted, or faulted before the erosion surface formed and an angular discordance is present between the rocks below and above the erosion surface, the unconformity is called an angular unconformity. If part of the section is missing as a result of nondeposition or erosion in an otherwise normal stratigraphic section, the name disconformity is applied to the erosion surface.

FIELD CHECKING YOUR INTERPRETATION

Once you begin to formulate ideas about the type of structural configuration you think is present in your map area, try to think of ways you can test those ideas. Generally, the tests consist of predicting what you will find in a particular place and then field checking to see if observations confirm your hypothesis. Often field checking will refine your interpretation even if it is generally correct. If your hypothesis is wrong, you must then begin to formulate a new hypothesis about the structure. Experienced mappers continually formulate and test ideas as the mapping proceeds.

PREPARING THE FINAL MAP

The following checklist gives some general guidelines for preparing your final map, Figure 3–6d. Your instructor may suggest which variations from these guidelines are suitable for your map. Check your final map to see that it conforms to these points:

1. The final map should consist of a base map (generally a topographic map) with a mylar (or tracing paper) overlay, showing the geology.
2. All of your line work should be done with permanent ink, or, if permitted, a hard pencil.
3. Use a tracing guide, dry transfer, or printout from a computer or other mechanical printing device for lettering.
4. The trace of contacts between rock units and faults must show on the map. Generally, some of these will be dotted, some dashed, and some solid (see definition of symbols). These lines should be drafted with a sharp, hard pencil or drawn in ink.
5. All map symbols, including lines used for contacts that appear on the map, should be defined in the explanation.

Computer graphics programs such as Canvas or Adobe Illustrator are now being used to prepare some geologic maps. These programs are capable of showing line work, lettering, symbols, and patterns. By scanning topographic maps, the geologic information can be superimposed on a topographic map base, and the final map can be printed in color. Digital maps have the advantage of being easily revised.

Final Map Checklist

_____ 1. **Title:** Generally, the title should include a geographic name for the area.

_____ 2. **Scale:** The map scale should be shown as a fraction (e.g., 1 : 62,500) and a bar scale. The bar scale should be calibrated in metric units as well as English units.

_____ 3. **North arrows:** Arrows showing both magnetic and true north should be on the map. Indicate the declination beneath the north arrows.

_____ 4. **Stratigraphic column:** Arrange this column with the youngest rock unit at the top and the oldest rock unit at the bottom. The column should consist of small boxes containing the symbol used on the map to identify the rock unit, and if the map is colored or if patterns are used, they should also appear in the box. Write the name of the rock unit next to the box.

_____ 5. **Author's name.**

_____ 6. **Strike and dip symbols:** These symbols indicate the orientation of rock units. Symbols should not be placed so close together that they overlap one another.

_____ 7. **Traces of the axes of folds:** These traces should show symbols to indicate the fold type.

_____ 8. **Traces of faults:** Fault traces should be shown with symbols to indicate the type of fault.

_____ 9. **Symbols used to identify formations:** The symbols should appear within each outcrop belt. It should be possible for someone who is unfamiliar with the map to determine the identification of the rocks found in every part of the map.

_____ 10. **Colors:** If colors are used on your map, they should be applied lightly. Use the side of a hard colored pencil. Be careful not to create streaks of color. Ideally, the color should be uniform. This may be obtained by rubbing the map with a soft piece of paper to help spread the color.

_____ 11. **Patterns:** If patterns are used on your map, select ones that will not make the map hard to read. Patterns may be applied freehand or from prepared sheets of symbols (called zip-a-tone). By convention, dots are used to represent sand and sandstones; dashes are used to represent shale; a brick pattern is used to indicate limestone; and small _vs_ or _xs_ are used to indicate igneous rocks.

_____ 12. **Cross sections:** Geologic maps are generally accompanied by geologic cross sections drawn along lines that cross the structure more or less at right angles to the trend of the structural features.

Your map should be legible, unambiguous, and complete, so someone who is unfamiliar with the area could use it.

Identification and Description of Sedimentary Rocks

After you know your location, identification of the rock types present at the outcrop is the next critical piece of information you need to collect. In general, sedimentary rocks are stratified and most are composed of grains; igneous rocks possess a distinctive texture described as an interlocking mosaic of crystals; and many metamorphic rocks exhibit a texture like that of igneous rocks, and fabric consisting of strongly aligned minerals or layering similar to that of sedimentary rocks. Unfortunately, you will find many exceptions to these generalizations. For this reason, it is advisable to give special attention to the distinguishing characteristics of those rocks that resemble one another. Some of these look-alikes are noted in the following rock descriptions.

FIELD DESCRIPTION OF SEDIMENTARY ROCKS

Until you achieve the level of expertise needed to confidently identify sedimentary rock units in your map area, it is advisable to record complete descriptions of features you observe. Be sure to measure and record the strike and dip of the bedding, and to note the presence of any fossils. Following are some other features you should observe.

Stratification

Describe the bedding (note that *bedding, layering,* and *stratification* are synonyms) of each unit exposed at the outcrop. Changes in color, composition, or texture of the material in the bed provide the basis for identifying individual beds. Geologists commonly describe beds as laminated, thin-bedded, or massive. Because these terms are subjective, you may find it useful to measure bed thickness. The weathering of beds is an important key to identification. Some thin or laminated beds may weather to form thick and massive ledges.

In examining strongly deformed sedimentary rocks, care must be exercised to distinguish bedding from **rock cleavage.** Shales and other fine-grained rocks commonly contain closely spaced planes, called cleavage, along which the rock may break. Cleavage is generally parallel or subparallel to the axial plane of folds formed in fine-grained sediment.

Composition

Most sedimentary rocks are mixtures of several components. For example, most limestones contain some clay or silt. You should learn to identify the mineral constituents of coarse-grained sediments, but the composition of fine-grained sediments may be more difficult to define. Remember that quartz has a hardness of 7 and will scratch metal, whereas calcium carbonate will effervesce in dilute hydrochloric acid. Also remember that mixtures containing either of these common components may exhibit the same characteristics. Thus calcareous shales will effervesce, as will powdered dolomite.

Key compositional terms

calcareous	—containing calcium carbonate
siliceous	—containing silicon dioxide (forms of quartz)
ferruginous	—containing iron
carbonaceous	—containing carbon
argillaceous (argillite)	—containing clay minerals

Texture

Grain size, shape, and fabric of sedimentary rocks define their texture. Most sedimentary rocks are either clastic (composed of fragments), chemical, or biochemical precipitates. The conditions under which sedimentation took place may strongly influence the texture. These conditions may be reflected in the primary sedimentary features present in the rock.

Key textural terms

arenaceous (arenite)	—containing sand
breccia	—containing angular fragments
clastic	—made up of fragments (e.g., clay, silt, sand, gravel)

Size categories of clastic materials

clay (lutite)	—particles less than 1/256mm in diameter
silt	—particles between 1/16mm and 1/256mm in diameter
sand	—particles between 2mm and 1/16mm in diameter
gravel	—particles between 4mm and 2mm in diameter
pebble	—particles between 64mm and 4mm in diameter
cobble	—particles between 256mm and 64mm in diameter
boulder	—fragments larger than 256mm in diameter
oolitic	—containing small, spherical masses formed by chemical precipitation of material deposited in concentric shells (If the spheres exceed 5mm in diameter, apply the term *pisolitic.*)
concretionary	—rounded masses of mineral matter formed during sedimentation or during the lithification of the rock.
stalactitic	—shaped like stalactites (commonly straw- or cone-shaped masses)

Color

Weathering commonly causes the surface of a rock to have a completely different color from the fresh unweathered material. These changes may prove to be characteristic features of the material and are useful in making identifications. Be sure to examine and describe the color of both the weathered and unweathered rock. For precise standard description of rock colors, use the rock color chart prepared by the Geological Society of America.

Sedimentary Rock Types

Special names are applied to varieties of sedimentary rocks that have certain distinctive compositions or textures. Many sedimentary rocks are mixtures. Thus, shales commonly contain some carbonate while sandstones contain some proportion of clay size particles, etc. The names assigned to these mixtures commonly use the key compositional terms defined earlier as modified to the predominant type of sediment (e.g., carbonaceous sandstone, or calcareous sandstone). Following are the distinctive features of the most common of the sedimentary rocks:

arkose	—clastic rocks containing a high percentage of feldspar
bioclastic limestone	—clastic rocks composed in large part of fossil fragments
breccia	—coarse, clastic rocks containing angular fragments
chalk	—limestone composed of microscopic-sized shells of protozoans
chert	—dense, cryptocrystalline (crystals are not visible without a microscope) siliceous deposit (also called flint)
conglomerate	—coarse, clastic rocks containing rounded fragments
coquina	—coarse, clastic rocks composed of a semiconsolidated mass of shell fragments, mud (commonly limy mud), or sand
dolomite	—a carbonate rock containing magnesium as well as calcium
graywacke	—clastic rocks, generally gray in color, containing small rock fragments, or dark minerals such as pyroxene or olivine
micrite	—dense, fine-grained (particles not visible to the naked eye) limestone
oolitic limestone	—limestone composed of small, spherical, egg-shaped bodies
sandstone	—any sedimentary rock composed of sand-sized (see preceding size categories) particles; composition should always be specified (e.g., quartz sandstone)
siltstone	—composed of silt-sized particles (see size categories)

ENVIRONMENTS OF DEPOSITION

Based on the type of stratification, composition, bedding, and primary sedimentary features present in the rock, you may be able to infer the type of environment in which the sediments form. If the conditions under which a sediment originated is known, it is appropriate to apply genetic terms as part of the rock description. The following are examples of terms applied to different types of depositional environments:

Nonmarine

eolian	—wind-blown dune deposits
fluvial	—stream channel and overbank deposits
glacial	—till and drift
lacustrine	—deposited in a lake
paludal	—low-lying coastal plain swamp and marsh deposits

Marine (deposited in the oceans or marginal seas)

supratidal	—above high tide
littoral	—intertidal
neritic	—on the continental shelf in water ranging from 10 meters to 200 meters deep
oceanic	—open ocean; continental slope and rise deposits
abyssal	—in water depths generally exceeding 4,000 meters

PRIMARY FEATURES IN SEDIMENTARY ROCKS

Primary structural features are those features that form contemporaneously with sedimentation. Stratification is the most common of the primary features formed during sedimentation. Some of these features such as ripple marks, raindrop impressions, and tracks may occur on bedding surfaces. Other features such as cross-bedding, graded bedding, and some fossils form within the layers.

Conditions of deposition determine what features form. Stratification is the most universal primary feature of sediments. Even the most uniform sediments usually possess some variations in texture, color, or composition that can be used to determine the bedding surface. Such surfaces usually indicate the interface between the sediment and the depositing medium. Beds may have been deformed subsequently by such processes as sliding, slump, plastic yielding during compaction, or even by folding and faulting in response to externally applied stresses. But bedding is important because it provides a system of reference planes. These serve to determine the extent and nature of any postdepositional deformation and make it possible to restore a sedimentary rock body to its original position during deposition. The type of stratification and its orientation may also provide valuable insight to the degree of agitation of the sediment by wave action; the existence, strength, and direction of currents, and even water depth. Primary structural features may consist of internal structure within the layers of the sediment, or they may be markings confined to the bedding surfaces. Many of these features (see ripple marks, swash marks, cross-bedding, and flow cast in the following sections) are directional and can be used to identify the top and bottom of the layers in which they occur, and all of them are valuable indicators of sedimentary environment. The determination of bed tops and bottoms is important because after deposition, beds may be folded and faulted so that bedding is vertical, overturned, or even completely inverted.

TEXTURAL VARIATIONS IN SEDIMENTARY ROCKS

One type of primary structure may be attributed to variations in mineral content or texture that arise from the deposition of different materials in a particular sequence. Examples of features of this type are salt sequences, graded beds, varves, and flat-pebble or sharp-stone conglomerates (Figures 4–1 and 4–2).

1. Salt sequences.   When salt water is evaporated, several salts are precipitated. They start coming out of solution in the following sequence:

a. Calcium carbonate and iron oxide

b. Calcium sulfide (gypsum and anhydrite)

c. Sodium chloride (salt)

d. Bittern (magnesium) salt

2. Varves.   Varves are rhythmically banded sequences of sediment in which a particular sequence of beds is repeatedly deposited. The most common seasonal varia-

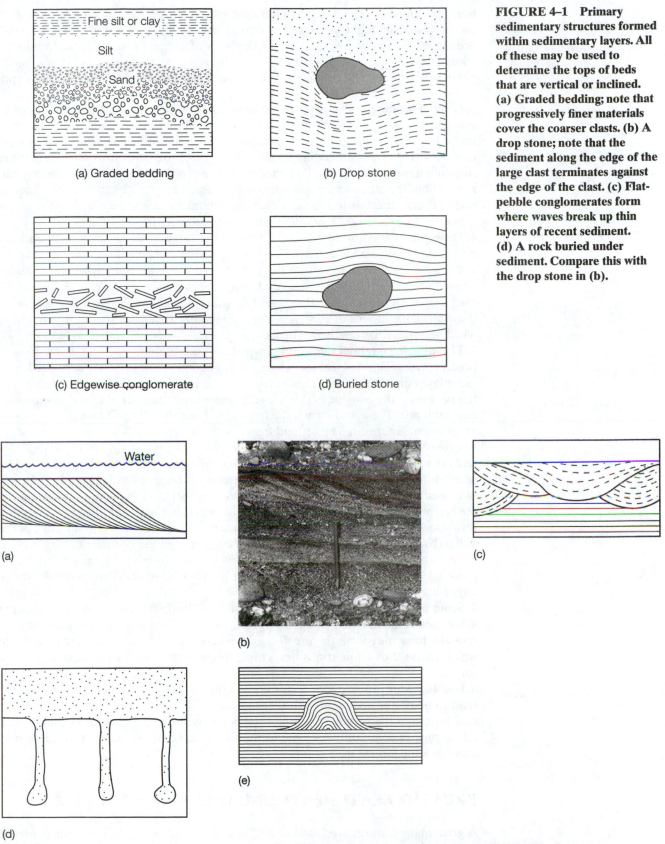

FIGURE 4–1 Primary sedimentary structures formed within sedimentary layers. All of these may be used to determine the tops of beds that are vertical or inclined. (a) Graded bedding; note that progressively finer materials cover the coarser clasts. (b) A drop stone; note that the sediment along the edge of the large clast terminates against the edge of the clast. (c) Flat-pebble conglomerates form where waves break up thin layers of recent sediment. (d) A rock buried under sediment. Compare this with the drop stone in (b).

(a) Graded bedding

(b) Drop stone

(c) Edgewise conglomerate

(d) Buried stone

Water

(a)

(b)

(c)

(d)

(e)

FIGURE 4–2 Primary sedimentary structures formed within sedimentary layers. (a and b) Cross-bedding. (c) Scour and fill cross-bedding. (d) A sketch of burrows that open upward. (e) A sketch of an algael mass in its growth positions.

tions are found in glacial varves where the lake is frozen over in winter, cutting off the supply of sediment, with the consequent development of a thin, often dark-colored layer. When the lake thaws, a thicker layer of sediment, newly brought into the lake, settles out during the summer. Varves can arise from seasonal variations in supply of sediment, type of sediment, or biological processes that affect sediment deposition.

3. Graded bedding.   This is the name given to sequences in which there is a systematic vertical change in sediment grain size. Usually a graded sequence consists of lower layers of larger grain sizes with progressive diminution in size going upward (Figure 4–1a). Graded bedding forms when a range of grain sizes of sediment are suspended and allowed to settle either in still water or in water with low-velocity current action. The currents tend to remove the finest sizes, which can be maintained in suspension or transported. Graded beds are commonly associated with the relatively sudden introduction of sediment into a particular environment.

4. Drop stones and buried gravel.   Occasionally, large isolated clasts occur encased within clay or silt-sized particles. These may provide valuable indicators of bed tops. If the large clast came to rest on the bottom and was later buried, fine laminations above the clasts will thin out and bend over the clast, as shown in Figure 4–1b. If the clast dropped into fine-grained sediment, it is likely to deform the laminations beneath the clast. Drop stones of this type may form where rocks are frozen into ice from which the rock drops as the ice melts.

5. Flat-pebble conglomerates and sharp-stone conglomerates.   Conglomerates or breccia layers (Figure 4–1c) form where wave action breaks a newly deposited sediment layer and moves the fragments a short distance from their point of origin. If the pebbles in the conglomerates are standing on end, the rock is called an **edgewise conglomerate.** These form when the fragments from the disrupted layer are trapped as sediment is deposited all around them.

6. Cross-bedding.   Consists of successive, systematic, internal beds or laminations that are inclined to the principal surface of accumulation. Cross-bedding occurs under many different natural circumstances as a result of transport and deposition of granular sediment in a current. **Deltaic cross-bedding** forms where fore-set beds are deposited on the front of a delta. It usually consists of three sets: (a) bottom-set beds that form from the deposition of fine sediment on the flat bottom beyond the edge of the delta; (b) foreset beds that are thicker and coarser than the bottom-set beds and inclined; and (c) top-set beds that are laid on top of the delta (Figures 4–2a and b), more or less flat lying. Cross beds formed by windblown sand are generally much larger than those formed under water.

7. Scour and fill.   This term applies to small-scale bottom scour or channels that are subsequently filled. When filling takes place, cross-bedding of a type known as **festoon cross-bedding** may form (Figure 4–2c). These cross beds have distinctive patterns in which each set of laminations lies within an erosional trough and have a concave surface.

8. Fossils.   May provide good indications of the top of beds. Shells such as clamshells tend to lie on the sediment with their concave side down. The burrows of animals that bore into the sediment on the sea floor may indicate bed tops (Figure 4–2d). Algae masses commonly take the forms shown in Figure 4–2e. Each of these may serve as a top indicator.

PRIMARY FEATURES FOUND ON BEDDING SURFACES

A great many features are found on bedding surfaces. Among these are ripple marks, mud cracks, swash and rill marks, impressions, pits and mounds, and scour and fill structures.

1. Ripple marks. These form as a result of movement of granular sediment in the medium of deposition. Ripples are common features in sand, both in sand dunes, where the sand is moved by air, and along a beach or in a stream, where the sand is moved by water. Many different types of ripples are found, but the most common ones are symmetric oscillation ripples and asymmetric current ripples. Oscillation ripples form as a result of wave action that induces movement in water to a depth equal to about half the wave length of the waves. When the water depth is less than that distance, the motion touches the bottom sediment and moves the sand and silt-sized particles back and forth along the bottom under the wave. The resulting ripples are symmetrical and have upward-pointed crests (Figure 4–3b). Current ripples form where water or air flows are directed across the sediment surface (Figure 4–3b). Sand is moved along by rolling or bouncing. As soon as irregularities develop in the surface, the character of the water or air movement on the upcurrent and downcurrent side of a crest is different. An eddying effect is created on the downcurrent side, which is protected from the movement of the type seen on the upcurrent side. Grains slip and roll down the downcurrent side, and the ripple develops an asymmetric profile (Figure 4–3b). The shape of such ripples cannot be used to determine bed tops, but sometimes the trough will accumulate coarser or heavier grains, indicating the top side of a layer. Current ripples formed in water can be distinguished from those formed in air by the ratio of wave length to wave height. Those formed in water have smaller ratios (e.g., 15 or less) than those formed in air (e.g., 15 or more). When currents move alternatively from two directions, an interference pattern may develop; other ripples are tongue-shaped.

2. Swash marks. Those patterns left in beach sand as a result of the forward rush of water after an incoming wave breaks are called **swash marks.** They are often concentric, arcuate patterns. As the waters return down the slope of the beach, they join to form a system of **rills** that feed into larger and larger flow channels. Other characteristic patterns are formed where rocks or shells interrupt the return flow of water.

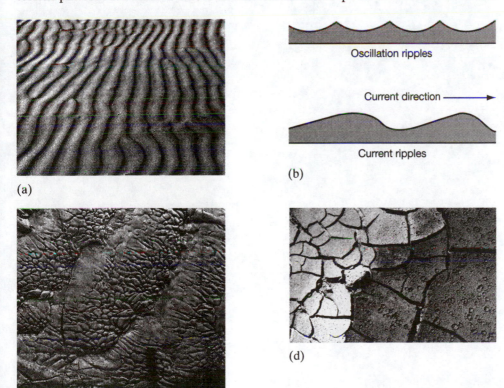

(a)

Oscillation ripples

Current direction ⟶

Current ripples

(b)

(d)

(c)

FIGURE 4–3   Primary sedimentary structures formed on bedding surfaces. (a) Ripple marks. (b) Profiles of current and oscillation ripple marks. (c) Load casts. (d) Mud cracks.

3. Impressions.   Many different types of processes and events leave marks or impressions on the sand or soft sediment on the sea floor. Animals leave **tracks, trails,** and sometimes burrows in the sediment. **Raindrop and hail imprints** with characteristic depressed center and raised edges are found. Drifting plants or other objects may **drag** objects across the sediment, leaving marks. **Bubbles** of gas may be held on the bottom in shallow water long enough for clay-sized particles to accumulate on the bubble. When the bubble breaks or is compressed, it leaves a characteristic circular impression on the sediment. Small **pit** and **mound** features may develop from the escape of gas from sediment. **Ice crystals** may leave crystalline patterns when they form on the sediment surface.

4. Flow casts or load casts.   Flow casts such as rolls, lobate ridges, or other raised features may be produced in a sandstone bed as a result of the flowage of sediment in an underlying bed, usually composed of some soft sediment (Figure 4–3c). Load casts may be bulbous, baglike, downward protrusions of sand produced by plastic yielding of an underlying, unevenly loaded soft sediment.

5. Mud cracks.   These are cracks that form most commonly in mixtures of clay, sand, and silt when the sediment is dried. The cracks usually form polygon-shaped patterns, and the mud layers frequently bend or curl upward (Figure 4–3d); the cracks are also wider at the top than at the bottom. Sand or silt sometimes fills the cracks when the cracks form in a mud layer that lies over a water-saturated sand layer.

Use of Aerial Photographs in Mapping

Photographs have the special advantage of portraying the land surface or rock features as they are seen by the eye. Aerial photographs are useful because they record detail not shown on topographic maps, and because they provide a different perspective. They may also provide more recent data than topographic maps, which are revised infrequently. Aerial photographs may be taken as oblique views of the land, sometimes showing the horizon (low-angle oblique) and sometimes with the camera pointed at a steep angle. Pictures have now been taken not only from planes flying at low altitudes but from planes flying fifty thousand or more feet high. Photographs are also taken from even greater altitudes by rockets and satellites.

Vertical Aerial Photography

Photographs taken with the camera pointed vertically down are a standard tool used in mapping (Figures 1–2a, 1–4a, and 5–1). These vertical aerial photographs are used in the production of topographic maps, and they make excellent bases on which many types of data may be recorded.

VERTICAL PHOTOGRAPHS

Vertical photographs are used to make most modern topographic maps, and they are used in place of, or to supplement, topographic maps for mapping of such things as soil, geology, land uses, and forest cover. While topographic maps provide elevation control data, aerial photographs can reveal the type of ground cover, rock outcrops, soil color, and many more cultural features than are shown on topographic maps. Vertical photographs find many uses in the fields of geology, geography, forestry, agriculture, urban and regional planning and development, environmental planning, and other types of land-use studies.

Extensive aerial photographic coverage for North America is available through the various governmental agencies, such as the U.S. Geological Survey, National Oceanic and Atmospheric Administration, the Coast and Geodetic Survey, and the U.S. Department of Agriculture (Forest Service and Soil Conservation Service), as well as from numerous private firms. The government agencies will supply photo index sheets, which show a rough mosaic of the photographs covering some specific area (i.e., a county or quadrangle). Both contact prints (9 × 9 inches) and enlargements are available.

Stereographic Photography

When vertical stereographic coverage is used for the preparation of modern topographic maps, overlapping photographs are taken from an altitude that will yield a photograph that is close in scale to the scale of the map. Much recent coverage has been taken at a scale of 1 : 24,000, the scale of 7½-minute quadrangles, but many other

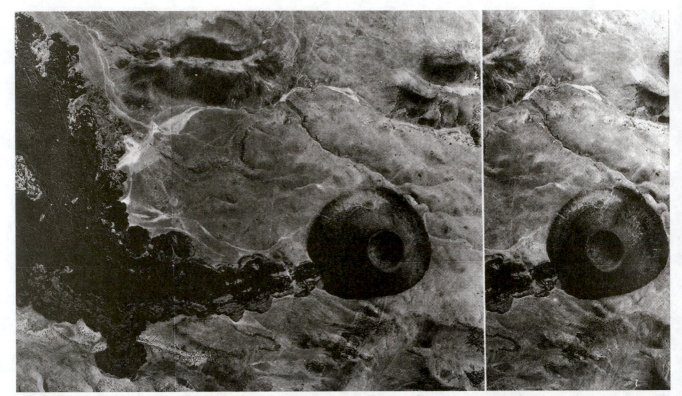

FIGURE 5–1   Stereopair of vertical aerial photographs of an area in Coconino County, Arizona. The scale is approximately 1:27,000. The photographs were taken from 9,300 feet. (Stereogram prepared by the Committee for Aerial Photography at the University of Illinois from photographs taken by the United States Department of Agriculture.)

scales are and have been used. The surest way to check the scale is to pick out two features in the central part of a photograph that appear on topographic maps of the same area, and then to measure and compare the distances, using the scale of the topographic map to determine the scale of the photograph.

Great care is taken to provide photographs of maximum usefulness. Normally, photographs are taken only under the best conditions—on cloudless days and in high mountains, when the sun is high in the sky, to avoid long shadows. Photographs taken for the purpose of making geologic maps are often taken during winter months, when the leaves and undergrowth do not obstruct outcrops; however, snow cover is avoided. Flight lines are kept as close as possible to a given altitude to provide uniform scale; the camera is pivoted so it remains vertical, and pictures are taken at intervals that will allow 60 percent overlap of two adjacent pictures in the same flight line; and the next flight line is positioned to allow a thirty percent overlap on the adjacent coverage. The overlapping photographs make it possible to obtain a three-dimensional view of the land surface (Figure 5–2). A device known as a stereoscope is generally used for this purpose.

Pocket-type stereoscopes are relatively inexpensive and are of great value in photo interpretation. Only those parts of the photographs that appear on both of the photographs can be seen in three dimensions. Placement of the photographs under the stereoscope is critical. Unless the photographs are correctly placed, you will be unable to see a three-dimensional image. Start by identifying some prominent landmark, such as a lake or mountain peak, on the photographs. Place the two photographs side by side so the landmark is directly under each of the lenses of the stereoscope. As you look through the stereoscope, move the photographs until the two images are superimposed and appear in three dimensions.

For best results, photographs from the same flight line should be used; illumination of the photographs should be arranged so light shines on the photo from the

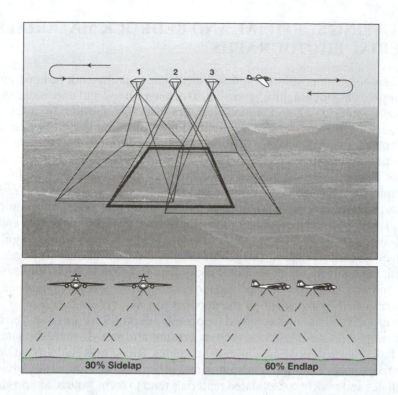

FIGURE 5–2   **Aerial photographs are taken in sequences that allow approximately a 60% overlap in the direction of the line of flight of the airplane. Adjacent flight paths are spaced so the photographs in one flight line overlap approximately 30% with those in the adjacent lines. (After Avery, T. E., and G. L. Berline, 1992,** *Fundamentals of Remote Sensing and Airphoto Interpretation,* **5th ed., Macmillan.)**

same direction in which the sunlight was shining on the ground surface when the photographs were taken. When the lighting is incorrect (sometimes even when it is correct), an inverted image is seen. Streams appear to flow along ridges, and high mountain peaks look like big holes. You will notice that relief on aerial photographs appears vertically exaggerated. The amount of this exaggeration depends mainly on the altitude of the plane and on the lenses in the camera, and is often about 2.5 times the actual relief. It will be useful for you to compare the relief as shown on a topographic map with what you see on the stereopairs.

With a little practice, some people can see three-dimensional images without the stereoscope, but this may cause eyestrain. The technique is to place a file card vertically between the two pictures. Look at the pictures from a comfortable reading position (8–10 inches), trying to focus your eyes on the floor, and the images will slowly merge.

EXERCISE 5–1   Interpretation of Aerial Photographs—Coconino County, Arizona

Refer to the aerial photograph shown in Figure 5–1.

1. Using a transparent overlay placed on the left image, trace the outline of

    a. The most recent cinder cone (the cinders are a dark-colored vesicular rock called scoria).

    b. The vent at the top of the cinder cone.

    c. The most recent lava flow (the flow is composed of basalt).

2. The remains of three older cinder cones can be seen on this photograph. Where are they located relative to the youngest cone?

_____

3. How can you distinguish the older cones from the younger one?

_____

_____

IDENTIFYING SURFICIAL AND BEDROCK MATERIALS ON AERIAL PHOTOGRAPHS

Subtle changes in composition of soil, sediment, and rock may not be apparent from aerial photographs, but usually it is possible to distinguish soil and unconsolidated sediment, sedimentary rocks, and crystalline rocks from one another. In making these identifications, it is necessary to consider the tonal color changes in the photograph, climatic conditions, physiographic features, resistance of various rock types to erosion and weathering, and characteristic structural features associated with various rock types (see Figure 5–3). Stratification of sedimentary rocks is a prominent feature on photographs when the layers are inclined and when the layers differ in their resistance to weathering. Igneous rocks are generally much more massive, have rounded outcrops, and may show intrusive relations to other rock types. The gneissosity and schistosity of some metamorphic rocks show up prominently in areas with thin soil cover. Mafic igneous rocks like basalt and gabbro are much darker in color than sialic rocks like granite and diorite. Also, the mafic rocks tend to break down in moist climates faster than sialic rocks. Recent glacial deposits and other unconsolidated sediments can be recognized by the shape of the deposits just as a recent lava flow or ash fall might be identified. Sinkholes, enclosed topographic depressions, are characteristic features developed on soluble rocks, such as gypsum and salt, or on carbonate rocks, such as limestone and dolomite, in moist climates. Ridge makers in moist climates are quartzites, well-cemented sandstones, and other insoluble materials while limestones, shales, and poorly consolidated materials tend to form valleys. In arid climates, limestones may also be ridge makers.

Geologists use several criteria to recognize the types of sediments or rocks that are shown on aerial photographs. Here are some of the most important criteria:

1. Intrusive igneous rocks.   Contacts between igneous rocks and other rock types commonly show intrusive relations (Figure 5–3c) where the igneous rocks cut across bedding of sedimentary rocks or foliation of metamorphic rocks. Granitic bodies tend to stand high, weather into rounded, massive forms, and are dome shaped. Vegetation is not normally abundant.

2. Extrusive igneous flows.   Basalt flows are generally distinctive (Figure 5–1). Sharp edges surrounded by talus piles are sometimes seen along margins of basalt flows. Blocky and ropy lava may sometimes be recognized. Because basalt tends to have low viscosity, thin flows can spread for long distances, and flows are usually porous, having channels through which surface waters are lost. Consequently, surface drainage patterns are not developed. Where basalt has been chemically weathered, plant growth is plentiful, but recent flows or flows in arid climates are usually barren. Rhyolite is much lighter in color than basalt, "sickle-like" curvilinear drainage is often noted, and viscous flow patterns may be present.

3. Pyroclastic rocks.   These rocks are easily confused with other types of sedimentary rocks. The finer ash and dust deposits, tuff, are easily eroded and are commonly deeply dissected, forming a rough, badlands-type topography.

4. Sedimentary rocks.   Sandstone, when composed of quartz and cemented by quartz, is resistant to weathering in moist climates and holds up ridges, forms mesa tops, and is usually easily recognized by its light color and its resistance (Figure 5–3b). Feldspar is not nearly as resistant under these conditions, so arkose and graywacke are not as likely to stand up the way sandstones and quartzites do. The surface drainage pattern is often poorly developed and may consist of widely spaced streams controlled by a rectangular fracture system if the units are nearly flat lying. Shales, like tuffs, tend to form a characteristic topography. The rock is soft and erodes easily, forming hills that are gently rounded unless, as in arid climates or in the absence of vegetation, gullies are prominent. Limestone is soluble and often shows karst topography in various

(a)

(b)

(c)

(d)

FIGURE 5–3   Some bedrock and surficial materials form patterns that are readily recognized on aerial photographs. (a) The sand dunes shown in this photograph are in North Africa. Most dunes are composed of quartz sand, but dunes may also form in sand-size particles of gypsum, shell fragments, or other materials. (Photograph by the United States Department of Defense.) (b) A cliff-forming sandstone layer forms the prominent cliffs in this photograph of Black Dragon Canyon, Utah. (Photograph by the United States Department of Agriculture.) (c) Granitic stock (light color) intruded into metasedimentary rock (dark color); a younger dike cuts across both the granite and metamorphic country rock. (Photograph by the Royal Canadian Air Force.) (d) Sand and gravel form large alluvial fans on the flanks of Death Valley, California. (Photograph by the United States Geological Survey.)

stages of development. Sinkholes, karst windows, blind valleys, and other such features are easily identified on photographs.

5. Metamorphic rocks.   Quartzite has the same characteristics as tightly cemented sandstones, and marble is similar in topographic expression to crystalline limestones. Highly developed foliation or schistosity in schists or gneisses may show up on photographs as a faint linear pattern. If deformation involving folding accompanied the metamorphism, these linear patterns exhibit the curved patterns associated with folds. Massive gneiss is indistinguishable from granite.

6. Surficial materials.   Photographs are especially valuable in the recognition of many types of surficial deposits. Distinctive landforms are commonly associated with many surficial deposits. Thus, the shape of the land may be a guide to recognizing the presence of certain types of deposits. Some examples are cited here.

a. **Sinkholes** contain distinctive deposits of soil or collapse breccias.

b. **Alluvial fans** composed of stream gravel are recognized by their shape and characteristic braided stream patterns (Figure 5–3d).

c. **Scars** are formed where the stream has moved back and forth across the valley, leaving oxbow lakes filled by sediment mark valleys containing meandering streams.

d. **Fields of sand dunes** are clearly visible on photographs (Figure 5–3a).

e. **Moraines** composed of mixtures of sand and gravel, called glacial till, can be identified on photographs of many recently glaciated regions.

f. **Landslide deposits** are more readily identified on photographs than on topographic maps.

g. **Swamps, mud flats, beaches,** and **other coastal landforms** can be recognized on photographs.

EXERCISE 5–2   Interpretation of Aerial Photographs—Owl Creek, Wyoming

Refer to Figure 5–4.

Scale: 1 : 48,500

Stereogram prepared by the University of Illinois Committee on Aerial Photography. Stereopairs of these photographs have been cut and arranged for ease of viewing. Use of a pocket stereoscope to view these photographs is desirable, but unnecessary to answer the following questions. Before answering questions, examine the photographs carefully, noting especially variations in gray shading and drainage patterns.

1. Can you identify ridges without using a stereoscope? _____

_____

2. If your answer to question 1 is yes, which direction do the ridges trend? North is at the top of the photograph.

_____

3. In what part of the photograph do you see a meandering stream the course of which is indicated by vegetation (north, east, west, or south)? _____

4. Is the bedrock in this area sedimentary, igneous, or metamorphic? _____

5. Explain the reason for your answer to question 4. _____

_____

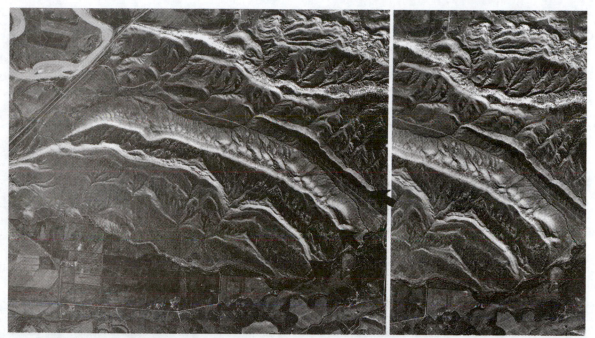

FIGURE 5–4   Stereopair of photographs at Owl Creek, Wyoming. (Stereogram prepared by the Committee for Aerial Photography at the University of Illinois from photographs taken by the United States Department of Agriculture.)

6. If the rocks are stratified, what is the approximate direction of strike? _____ If you are unfamiliar with the terms *strike* and *dip*, refer to Figure 3–2.

7. If the rocks are stratified, what is the approximate direction of dip? _____

8. What characteristic of the topography, and especially the shape of some ridges, allowed you to answer question 7? _____

    _____

9. Are all of the rock units in this area equally resistant to erosion? _____

10. Trace the contacts indicated by arrows across the photograph.

EXERCISE 5–3   Interpretation of Aerial Photographs—Lookout Ridge, Alaska

Refer to Figure 5–5.
Stereogram prepared by the University of Illinois Committee on Aerial Photography.

1. What is the strike of the light-colored units shown on this photograph? _____

2. What is the direction of dip of the strata exposed in this area? _____

3. Which of the rock units are most resistant to erosion? _____

4. What causes the sharp breaks in the continuity of the ridges west of the water gap?

    _____

    _____

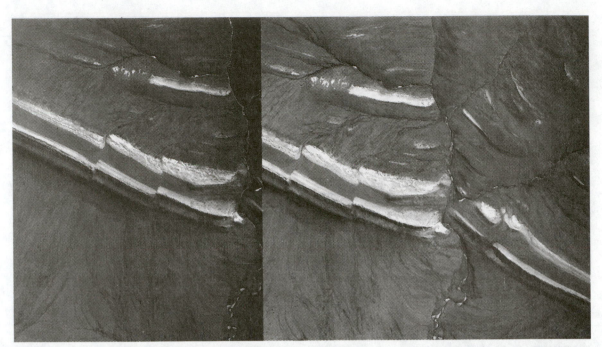

FIGURE 5–5   Stereopair of photographs of Lookout Ridge, Alaska. (Stereogram prepared by the Committee for Aerial Photography at the University of Illinois from photographs taken by the United States Department of Agriculture.)

Interpretation of Surficial Geologic Maps

Geologic maps of surficial materials show mainly unconsolidated sediment, such as alluvium, beach sands, lake deposits, sand dunes, talus, colluvium, etc. Bedrock maps show the mapper's interpretation of the distribution of bedrock map units beneath surficial materials, as well as where they are exposed at the ground surface. Maps that combine the two generally indicate where bedrock units are exposed and where they are covered only by soil. Surficial deposits are shown in other areas.

Surficial geologic maps should not be confused with soil maps. The Department of Agriculture and other government agencies prepare maps showing the distribution of soil types. The map units used on these maps are unlike those on geologic maps. The map units on surficial geologic maps are defined on the origin of material (e.g., stream alluvium, lake bed deposits). Soil maps have identification of soil fertility as a primary function. The two are related, but the names used on soil maps are quite different from those used on geologic maps.

USES OF SURFICIAL GEOLOGIC MAPS

Surficial geologic maps depict the distribution of natural materials that lie on top of bedrock or, in some areas, regionally extensive layers of semiconsolidated sediment. The term *bedrock* refers to hard, consolidated, and cemented rock that may be of igneous, metamorphic, or sedimentary origin. In some areas, semiconsolidated or even unconsolidated sediment lies beneath the soil and other surficial materials. For example, in the Atlantic and Gulf coastal plains, surficial materials such as stream deposits lie on marine strata that remain semiconsolidated even after many millions of years.

Most surficial materials seen on aerial photographs are unconsolidated or semiconsolidated sediments deposited on land or along the seashore. Some such deposits are thick; but others, such as soil, form a thin veneer on consolidated bedrock.

Maps of surficial deposits are especially important for those involved in construction or land-use analysis. Because the intended uses of maps of surficial materials differ so greatly from the uses of bedrock maps, it is useful to have bedrock and surficial materials on separate maps of the same area. Maps intended to illustrate bedrock commonly show surficial materials only where they are so thick or extensive that they obscure the bedrock to such an extent that the mapper cannot determine what lies beneath the surficial material. Note that many of the maps in the appendix show very limited areas of surficial material. Maps of surficial materials place emphasis on what is exposed at the ground surface, and in some instances all bedrock is mapped as a single unit.

TYPES OF UNCONSOLIDATED MATERIALS SHOWN ON GEOLOGIC MAPS

The following are some types of surficial materials that may appear on geologic maps. Refer to a text on geomorphology or sedimentology for information concerning the shapes and composition of these materials.

1. Stream deposits
    Alluvium (stream deposits)
    Alluvial fan deposits (deposits formed where streams issue from mountains)
    Floodplain deposits
2. Glacial deposits
    Glaciofluvial deposits
    Moraines (terminal, recessional, lateral, ground)
    Kame terraces (formed between edges of valley glaciers and mountain valleys)
3. Coastal deposits
    Tidal and mud-flat deposits
    Deltas
    Marine swamps
    Reefs
    Beaches
4. Slope deposits
    Talus (angular rock fragments that have moved down steep slopes)
    Mud flows
    Landslides
    Colluvium (mixtures of soil, alluvium, and other slope deposits)
5. Wind-laid deposits
    Sand dunes
    Loess (windblown dust and silt)
    Volcanic ash
    Welded volcanic tuffs (volcanic ash in which particles are fused together)
6. Spring deposits
    Tufa
    Geyserite (siliceous deposits formed around geysers)
7. Lake deposits
    Playa lake deposits (salts deposited in temporary lakes in desert basins)

EXERCISE 6–1   Surficial Deposits

Refer to the maps in the appendix (pages Ap2–13, 5, 25, and 27).
Identify the types of surficial deposits shown on each of the following maps, and describe how their distribution is related to the topography.

1. Grand Canyon: _____
2. Atkinson Creek: _____
3. Paradise Peak: _____
4. Salem: _____

EXERCISE 6–2   Surficial Deposits—Ouray Quadrangle, Colorado

Refer to the map in the appendix (page Ap2–23).

1. Trace the surficial deposits on an overlay. Label and identify each of the surficial deposits (use the legend). Examine the relationship between the deposits and the topography.

2. Which of the surficial deposits lie in the bottom of the valley in which Ouray is located? _____

3. Which of the surficial deposits lie on steep slopes? _____

4. How might you recognize the Qal and Qf deposits by examining the relationship between the deposits and the topography?

   _____

5. How do Qf and Qt differ in terms of their position in the topography?

   _____

EXERCISE 6–3   Surficial Deposits—Davis Mesa Quadrangle, Colorado

Refer to the map in the appendix (page Ap2–5).

1. Were all of the alluvial deposits shown on this map deposited from the Dolores River?

   _____

2. Cite evidence for your answer to question 1. _____

   _____

   _____

3. Where was the source of the alluvium shown in section 4? _____

   _____

4. Cite evidence for your answer to question 3. _____

   _____

EXERCISE 6–4   Surficial Deposits, Montvale Quadrangle, Virginia

Refer to the sketch of a portion of the Montvale Quadrangle, Figure 6–1.

Villamont-Montvale Quadrangle
Publication 35, 1981, Virginia Division of Mineral Resources
Geology, by W. S. Henika

Note that talus deposits (unlabeled) occur as scattered patches. The Blue Ridge fault is a subhorizontal fault that places Precambrian and lower Cambrian rock units on top of Cambrian shale.

   Stratigraphic Units That Occur in This Region

   Qal   alluvium
   Qtd   terrace deposits

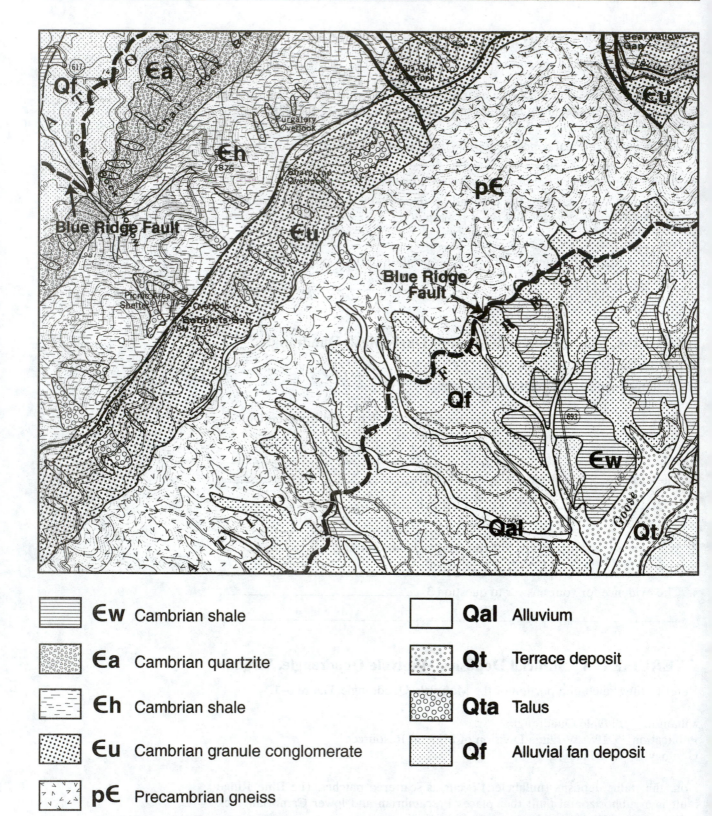

Єw Cambrian shale		**Qal** Alluvium
Єa Cambrian quartzite		**Qt** Terrace deposit
Єh Cambrian shale		**Qta** Talus
Єu Cambrian granule conglomerate		**Qf** Alluvial fan deposit
pЄ Precambrian gneiss		

FIGURE 6–1   Geologic map of part of the Montvale, Virginia quadrangle showing both bedrock and surficial deposits. (After W. S. Henika, "Geology of the Villamont and Montvale Quadrangles." Printed with permission of the Commonwealth of Virginia from Division of Mineral Resources Publication 35, Henika, 1991.)

Qf	fan deposits
Qta	talus
€e	Elbrook formation (dolomite and shale)
€w	Waynesboro formation (mainly shale)
€s	Shady formation (dolomite)
€a	Antietam quartzite
€h	Harpers formation (shale and graywacke)
€u	Unicoi formation (quartzite, sandstone)
€uL	Lower Unicoi formation
p€	gneisses

1. What is the apparent relative age of Qal and Qf? _____

2. Cite evidence for your answer to question 1. _____

3. What is talus? _____

4. Based on the resistance of the different rock units to weathering and erosion, which of these rock units do you think form ridges? (This area is located in a warm humid region.)

   _____

5. What is the probable composition of the talus? _____

6. Where do you think the talus is located in the topography? _____

7. Where would you expect the alluvial fan deposits to be located in the topography?

   _____

8. Using a transparent overlay, trace the contacts of the rock units beneath the surficial deposits.

   _____

Introduction to Geologic Maps of Bedrock

Ideally, geologic maps show the actual distribution of map units. Several factors limit the realization of this ideal. For maps of surficial materials, a limitation may be the accuracy with which the mapper identifies the materials in the field or interprets them from aerial photographs or images. In the case of maps of bedrock, the extent of exposure and the character of the map units are factors. Natural rock units of suitable thickness for mapping at the intended scale, and composed of an easily identifiable rock type, are the exception rather than the rule. This frequently means that the identification of map units is not entirely objective. Typically, large portions of the map area reveal no outcrops. Consequently, outcrop information must be extrapolated to cover the entire map. This extrapolation is based on an understanding of the structure and stratigraphy of the area. Mapping, then, proceeds by observation, hypothesis, testing, further observation, and so on. The final product is a mixture of recorded observations and the hypotheses that the mapper considers best to connect those observations.

Although geologic maps appear to represent complete information about the areal distribution and geologic relations of the underlying rocks, they are, in fact, based on a limited number of outcrop control points. When the geologic map covers a small area (i.e., a 1:24,000, 1:25,000, or 1:50,000 quadrangle map), the sedimentary rocks are usually subdivided into smaller rock units.

Geologic maps may depict the distribution of rocks of certain types or of certain ages. Commonly, the map depicts a combination of these two characteristics of crustal materials. When the map covers a large area (i.e., a state or country), the map is likely to show all sedimentary rocks deposited during a certain period of geologic time with the same color or pattern. Igneous or metamorphic rocks of that age may be differentiated by a different color or pattern. This system is used on the geologic map of the United States, published by the U.S. Geological Survey.

Given an unfamiliar geologic map, you should begin interpretation by examining the information given around the margins of the map. This will generally include the following:

1. Location. Small inset maps showing the location of the map may be present. Otherwise, the name of the area and the longitude and latitude will indicate its location. From its location, those familiar with regional geology will know what age rocks and surficial materials to expect, and what types of major structural features may be present.

2. Scale. The map scale provides a guide to the size of the area covered by the map and the size of features shown on the map. Maps have a bar scale or a fractional scale that indicates the ratio of a unit of horizontal distance across the map relative to the corresponding horizontal distance across the ground (e.g., 1:50,000).

3. Map Units. A stratigraphic column is always shown on geologic maps, and it is the most important source of information about the map. From the explanation, you can determine exactly what was differentiated on the ground. The choice of map units will indicate to what extent bedrock and unconsolidated surficial materials have been included on the map. Generally, the stratigraphic column is arranged with the youngest material or rock unit shown at the top; other units are arranged in order of increasing age. The geologic period during which the map unit was formed is indicated; and the color, map pattern, and abbreviations used to identify each map unit on the map are shown.

4. Symbols.   Geologic maps may contain a great variety of symbols. A chart showing the most commonly used symbols is provided on the inside back cover of this book.

After initial inspection, interpretation of the map will generally proceed in different ways, depending on the basic purpose for which the map was prepared, the intended use, and the type of information it contains. There are many specialized maps, but most "standard" geologic maps are designed primarily to illustrate unconsolidated materials found at the earth's surface, bedrock geology, or combinations of these two. Maps showing surficial materials are especially important for planners, engineers, and others whose primary need is for information about near-surface materials. In recent years, new types of geologic maps have been designed to emphasize the location of various types of natural hazards or areas that are environmentally sensitive.

Maps that depict both bedrock and surficial materials are much easier to interpret if the surficial materials are removed from the map and considered separately. This "stripping" of the surficial materials may be done mentally, but it is helpful to place an overlay on the map and trace the contacts of bedrock units, faults, etc. In making this tracing, the contacts between beds and faults are projected under the surficial materials. In areas of relatively simple bedrock structure, projecting contacts is easily done, but where the geologic structure is complex or the surficial materials are extensive, considerable uncertainty may exist.

PRIMARY SHAPE OF SEDIMENTARY ROCK BODIES

We tend to think of "ideal" sedimentary rock bodies as being layers that are uniform in composition and thickness and of great lateral extent. Some "real" rock bodies come close to having this shape. From the rim of the Grand Canyon, layer after layer of rocks in the canyon walls appear to extend with nearly uniform thickness from horizon to horizon. For many rock units, the ideal model of a uniformly thick plate is valid over small areas (e.g., across a 1:24,000 or 1:50,000 quadrangle), but if traced over great distances, even these layers vary in thickness. Almost all sedimentary units either thin out in such a way that they are wedge-shaped along their margins, or they gradually change composition and interfinger with other rock types laterally (Figure 7–1a, b, and c). Sedimentary rock units that were deposited on continental shelves or platforms and those laid down on the deep sea floor come closest in shape to the "ideal" model. Sedimentary rock units deposited on unstable continental margins vary greatly in thickness and may undergo rapid lateral changes. Where lateral changes in lithology and thickness occur over short distances, it is often difficult to define and map the rock units.

A number of conditions during sedimentation may cause layer thicknesses to vary. Sediments tend to accumulate in areas that subside (Figure 7–1a, b, and c). Thus sediment accumulation is commonly thicker where subsidence is rapid or continues for a long time. Strata may also increase in thickness toward the source area, as in the cases of fault-bounded sedimentary basins (Figure 7–1d) or intermontane basins. Layers of sediment eroded from the Rocky Mountains are thicker close to the mountains than they are far out in the Great Plains. During the Paleozoic era, large basins and domes were present on the North American Craton. Today, we find much thicker units in basins such as the Michigan Basin and the Appalachian Basin than

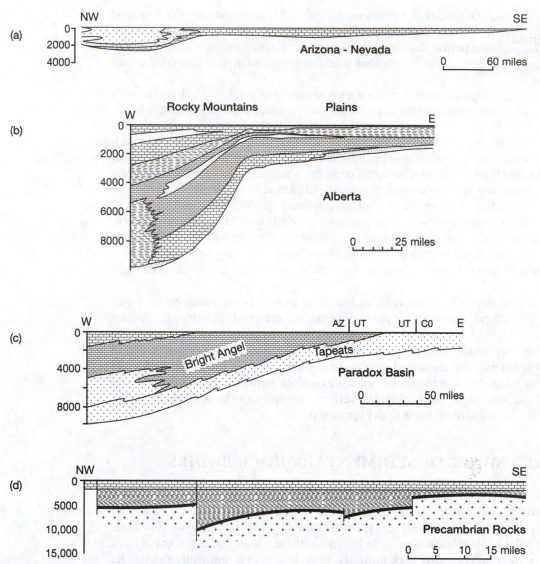

FIGURE 7–1   Examples of the lateral variations in the thickness and composition of sedimentary layers. (a) Cross section from Arizona to Nevada of Mississippian age Redwall Limestone and its equivalents. Note that the Redwall thins out to the southeast and changes into other rock units toward the northwest. (b) Dramatic changes in thickness and composition commonly occur where sedimentary units are traced from stable regions such as the craton (the plains) into unstable regions as interpreted by Douglas, et al., 1970. (c) Changes similar to those in (b) were mapped by Boars (1958) in the Paradox Basin region of the Colorado Plateau. (Boars, 1958.) (d) Connelly (1957) found sharp changes in thickness are associated with faults that were active during deposition of the Cambrian units in Kentucky. (Modified from figures in Cook, T. D., and A. W. Bally, Eds. © 1975, *Stratigraphic Atlas of North and Central America,* Princeton University Press. Courtesy of Shell Oil Company, © 1975.)

in surrounding areas on the craton. Rock units tend to thin out toward areas such as the Ozark and Adirondack domes that remained topographically high over long periods of geologic time.

Framework of sedimentary rock bodies	Characteristics of strata
1. Shallow-water deposits laid down over cratons and continental margins during periods of crustal stability.	1. Strata tend to be uniform in thickness, but change in composition laterally.

2. Basins develop within cratons during periods of crustal instability.

2. Strata gradually thicken toward the center of the basin.

3. Passive continental margins.
   a. shelf deposits

3a. Strata tend to thin toward shore where they may interfinger with beach deposits.

   b. slope deposits

3b. Strata are inclined down slope, and interfinger with shelf sediments.

4. Active continental margins.

4. Strata exhibit rapid changes in thickness and in composition.

5. Fault-bounded depressions (grabens).

5. Strata are wedge shaped, and thickness increases toward fault margin.

6. Deltaic deposits.

6. Strata are inclined on delta front.

7. Intermontane basins.

7. Strata increase in thickness toward mountain fronts.

The true thickness of a rock unit is the distance from the top to the bottom of the unit, measured perpendicular to the top and bottom contacts of the unit. Because unit thicknesses commonly vary, it may be useful to compile a map showing the geographic distribution of thickness. This may be depicted by drawing a contour map on which the contours are lines connecting points of equal thickness. Such a map is called an **isopach map** (Figure 7–2). The contours on an isopach map connect points of equal "true" thickness. In subsurface studies, much of the data are obtained from drill holes. Because most wells are vertical and layers vary in dip, the thickness of a unit penetrated in a well may not be the true thickness of the unit. Because of this effect, an **isochore map,** similar to an isopach map but based on the thickness of units as penetrated in wells, is sometimes drawn.

EXERCISE 7–1   Kilauea Crater, Hawaii

Refer to the map in the appendix, page Ap2–19.

The red lines on the right side of Kilauea Crater are isopach lines showing the thickness of the ash and lapilli deposits erupted in 1959. What was the prevailing wind direction during the 1959 eruption? _____

Lateral changes in the composition of sediments that were deposited during a given interval of time are represented on a special type of map called a **lithofacies map,** Figure 7–3. Such maps make it possible for geologists to reconstruct the types of sedimentary environments that existed during the selected time interval. Many strata contain more localized bodies of sediment that formed within layers of more regional extent. Bars of sand that formed offshore from beaches, channel fill, and reefs are examples of deposits formed in special environments. Because their porosity and permeability are commonly greater than that of the sediment that surrounds them, such deposits may contain economically important accumulations of oil, gas, or groundwater.

Although many formations are not uniform in thickness, most variations in thickness are gradual. Consequently, over areas the size of a quadrangle, layer

FIGURE 7–2   Map of the restored thickness and lithofacies of Lower Cambrian sandstones in the Rocky Mountains. In making a map of this type, the effects of later deformation and erosion are removed. The sandstones of Lower Cambrian age carry different names in different parts of the region. The Tapeats Sandstone of Arizona is equivalent to the Tintic Quartzite of Utah and the Bingman Quartzite of Idaho. The sandstones thin out toward the Great Plains, and data are not sufficient to complete the map to the west. (After Christina Lockman-Balk, in *Geologic Atlas of the Rocky Mountain Region,* **1972, Rocky Mountain Association of Geologists, Denver, CO.)**

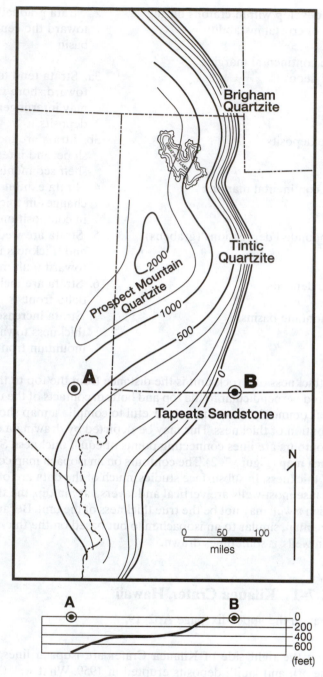

300X vertical exaggeration

thickness is essentially uniform. In describing the shape of sedimentary rock bodies in this book, it is assumed that it is possible to clearly identify the top and the bottom of the unit. In such cases, the shape of a rock unit may be defined by determining the shape of its contacts. Three sources of information about the shape of a contact may be available on geologic maps:

1. Strike and dip of layering (see page 32)
2. Structure contours
3. The geometric relationship between contacts and topographic contours

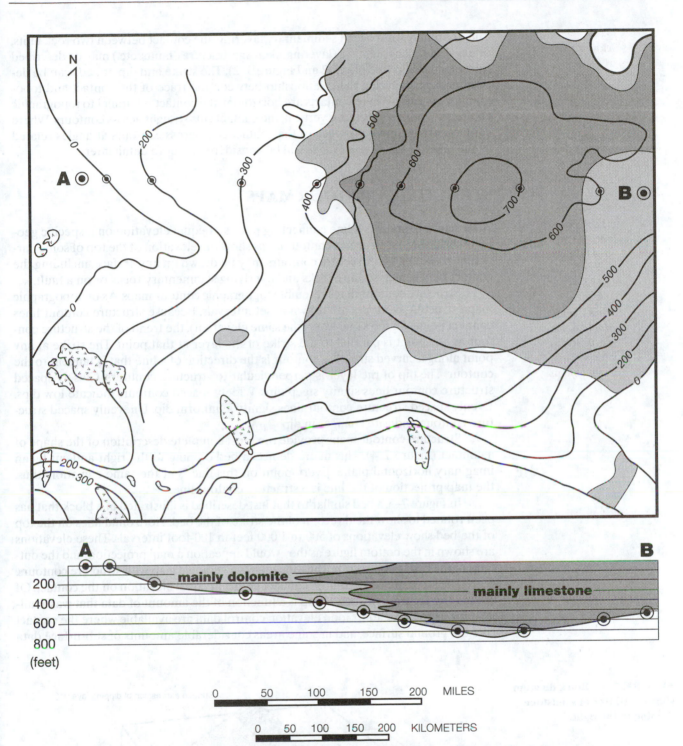

FIGURE 7–3 Isopach and lithofacies map of Mississippian rocks in a portion of the Williston Basin located in the Dakotas and Alberta. The lithofacies map indicates variation in the proportion of the carbonate that is dolomite. The section is nearly 700 feet thick in the southeastern part of this map area. The stippled areas are outcrops of Precambrian basement rocks in the Black Hills (right), the Bighorn Mountains (center), and the Beartooth Mountains (left). (After N. H. Foster in *Geologic Atlas of the Rocky Mountain Region,* 1972. Rocky Mountain Association of Geologists, Denver, CO.)

The orientation in space of sedimentary layering, the contact between two rock units, or any other planar feature (layering, cleavage, fractures, faults, etc.) may be described by strike and dip (see page 32 and Figure 3–2). The strike and dip of beds can be determined by analyzing the relationship between the trace of the contact and topographic contours. Where contacts are horizontal, the contact is parallel to topographic contours. Where contacts are vertical, the contact cuts straight across contours. Where contacts are inclined, the trace of the contact cuts across contours at angles related to the angle of dip. These effects will be considered in more detail later.

STRUCTURE CONTOUR MAPS

A **structure contour** is a line connecting points of equal elevation on a specific geologic surface. Most structure contour maps show the elevation of the top of some particular rock unit, but structure contours may be drawn on any surface, including the contact between crystalline rocks and overlying sedimentary rocks or on a fault.

Structure contour maps resemble topographic contour maps. As on topographic maps, structure contours are drawn at set intervals. Because structure contour lines connect points on the same level (the same elevation), the trend of the structure contour at any point is parallel to the strike of the layer at that point. The strike at any point along a curved structure contour is the direction of a line that is tangent to the contour. The dip of the layer is perpendicular to structure contours. Closely spaced structure contour lines signify steep dips. Widely spaced contours indicate low dips. Evenly spaced structure contour lines signify a uniform dip. Unevenly spaced structure contours indicate changes in dip.

Structure contour maps provide the most complete description of the shape of a contact (Figure 7–4). This figure depicts a bed dipping to the right and cut by an imaginary horizontal plane. Every point on this line is at the same elevation; thus, the map projection of the line is a structure contour line.

In Figure 7–5, a bed similar to that just described is illustrated in a block that has been rotated to show the direction of the strike of the bed. Horizontal lines on the top of the bed show elevations of 500 to 1,000 feet in 100-foot intervals. These elevations are shown in the bottom figure as they would appear on a map projection, and the outcrop of the bed is also shown. Thus, the map is a geologic map with structure contours.

The structure contour lines connect points of equal elevation on the contact. Of course, the accuracy of such a map is a function of the amount of data that was available for its construction. Good elevation control data are available where the contact cuts the ground surface, and in some areas considerable amounts of subsurface data

FIGURE 7–4 Block diagram showing a layer of sandstone dipping to the right.

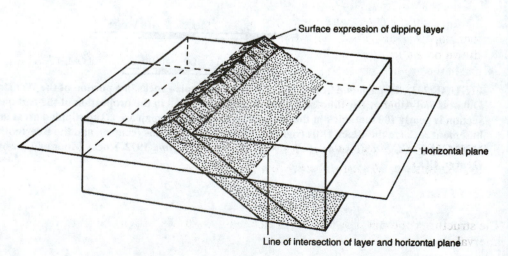

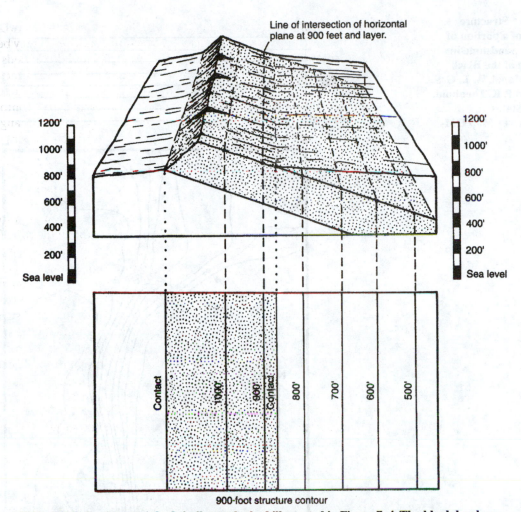

FIGURE 7–5   Above: A bed similar to the bed illustrated in Figure 7–4. The block has been rotated so we look in the direction of the strike of the bed. Horizontal lines (structure contours) on the top of the bed show elevations of 500 and 1,000 feet in 100-foot intervals. Below: These structure contours are shown as they would appear on a map projection, and the outcrop of the bed is also shown. The map is a geologic map with structure contours.

may also be available from wells, seismic lines, or other geophysical studies. Where data are adequate, structure contours are sometimes drawn on geologic maps. Several of the maps in the reference set in the appendix contain structure contours.

Profiles of the contoured surface can be prepared along any line on a structure contour map using the same technique used in drawing topographic profiles. Such a profile drawn on a structure contour map shows the structure (shape) of the layer along the line of the profile. It shows the elevation of points on the profiled surface. Only rarely are these also points on the ground surface. In Chapter 10, we will study examples of structure contour maps drawn on layers with different types of structural features.

EXERCISE 7–2   Structure Contours on the Flanks of the Black Hills, Wyoming, Montana, and South Dakota

Refer to Figure 7–6.

The structure contours shown in the sketch map are drawn with a 500-foot contour interval.

FIGURE 7–6   Structure contour map of a portion of the Fall River Sandstone in the area north of the Black Hills. (After Mapel, W. J., G. S. Robinson, and P. K. Theobold, 1959. United States Geological Survey Map OM-191.)

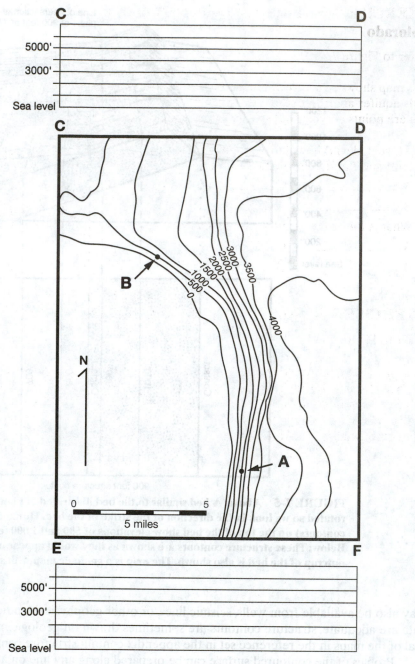

1. What is the direction of the strike of the bed on which the structure contours are drawn?

   a. Near the letter (A)? _____

   b. Near the letter (B)? _____

2. Draw a cross section from east to west across the top and bottom of the sketch map. Use the vertical exaggerations shown in Figure 7–6.

3. Describe the structure of the bed on which the contours are drawn.

   _____

   _____

   _____

EXERCISE 7–3   Structure Contour Map of the Denver Basin, Colorado

Refer to Figure 7–7.

This map shows structure contours drawn on the contact between the Laramie–Fox Hills aquifer and the overlying Laramie Formation, which is a shale unit. The circles are points at which subsurface data were obtained.

1. At what elevation would a well penetrate the contact between the Laramie–Fox Hills and Laramie formations at the points marked?

   A. _____   C. _____

   B. _____   D. _____

2. What is the strike of the Laramie–Fox Hills aquifer at the points marked?

   A. _____   C. _____

   B. _____   D. _____

EXERCISE 7–4   Isopach Map of the Lower Ordovician

Refer to Figure 7–8.

This map shows thickness measurements of the Lower Ordovician in the region of the Appalachian Basin. This part of the stratigraphic section is absent over the Waverly Arch in Ohio and Kentucky. The Lower Ordovician is also absent in the region east of the Great Valley of Virginia, where older rocks are exposed. The Lower Ordovician is about 1,000 feet thick along its eastern edge.

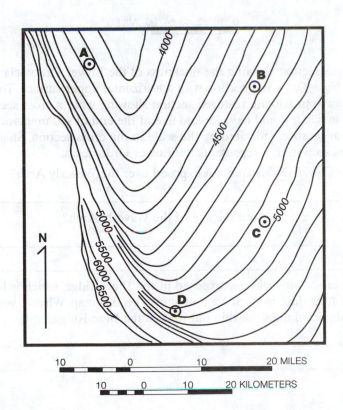

FIGURE 7–7   Structure contour map of a portion of the Denver Basin. Contours are drawn on top of the Laramie–Fox Hills aquifer. Contours are drawn at 100-foot intervals. The datum is mean sea level. (After Romero, J. C., and E. R. Hampton, 1974. United States Geological Survey Map I-791.)

FIGURE 7–8   Isopach map showing thickness of the Lower Ordovician rocks (mainly the Beekmantown Formation) in the area of Ohio, Pennsylvania, West Virginia, Kentucky, and Virginia. (Modified from H. P. Woodward, "Southeastern Appalachian Interior Plateau" AAPG Bulletin, Vol. 45, no. 10, AAPG © 1961, reprinted by permission of the American Association of Petroleum Geologists.)

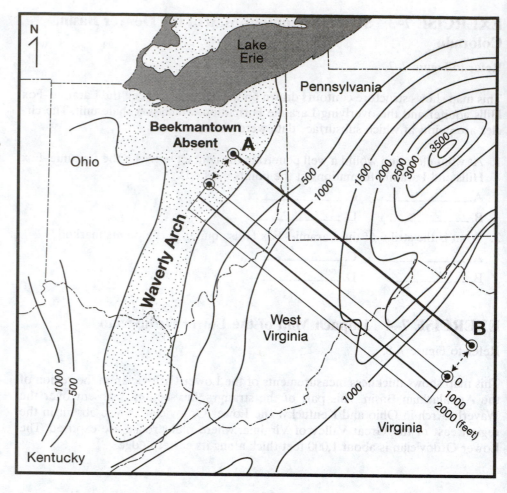

1. Draw a cross section showing the thickness of the Lower Ordovician along the line A–B. *Note:* Draw this section with a horizontal upper surface. The resulting section is an approximate restored section showing what a cross section of the Lower Ordovician would have looked like at the end of sedimentation. Use the vertical exaggeration indicated by the scale in the cross section. Shade the part of the cross section that represents the Lower Ordovician.

2. Why is the Lower Ordovician not exposed over the Waverly Arch?

_____

3. What does this tell us about the age of the Waverly Arch?

_____

_____

4. Today, Precambrian rocks are exposed in the Blue Ridge, which is located just east of the 1,000-foot isopach contour shown on this map. Why do you think the Lower Ordovician does not thin out toward the Blue Ridge?

_____

_____

Geologic Maps
of Homoclinal Beds

The term **homoclinal** applies to all contacts of uniform dip, whether layers are horizontal, vertical, or inclined layers. The term **homocline** is sometimes confused with **monocline**. **Homocline** refers to a single dip; **monocline** refers to a flexure (Figure 8–1). Only rarely do layers remain perfectly planar for great distances, but many layers are nearly planar, and the maps of such layers closely resemble those that are plane. The exercises in this section will help bring out these subtle differences.

While perfectly plane beds are the exception, beds that are only slightly warped or curved are widespread in North America. Beds fitting this description cover vast areas in the Great Plains, in the Atlantic and Gulf coastal plains, and in the Appalachian and Colorado plateau regions.

In some regions, such as the Colorado Plateau near the Grand Canyon, the beds are so nearly flat that careful examination of the changes in elevation of contacts over long distances is needed to detect the dip. In such cases, structure contour maps are especially valuable as a way of illustrating the tilt and shape of the surface. For this reason, geologic maps with structure contours are used in the following exercises. Remember that structure contours are drawn on only one bed. Thus if beds change thickness, the shape of surfaces higher or lower in the section will be different.

PATTERNS OF HOMOCLINAL BEDS ON GEOLOGIC MAPS

Because their geometry is so simple, it is easy to understand the appearance of homoclinal contacts on maps (Figure 8–2). Complexities in the map patterns of homoclinal contacts are due mainly to the irregularities in the topography. Even these need pose little difficulty if the following characteristics of such beds are kept in mind:

1. The contacts of horizontal layers are parallel to topographic contours.
2. Contacts of vertical layers form straight lines on maps regardless of the topography.
3. Contacts of dipping beds form V-shaped patterns where they cross ridges or stream valleys.

V-SHAPED CONTACT PATTERNS ON GEOLOGIC MAPS

On detailed geologic maps at scales of 1:100,000 or more, contacts between rock units generally have the shape of a V where a stream crosses the contact (Figure 8–3). If the V-shaped pattern occurs in a stream valley, the V generally points in the direction of dip of the layers, regardless of whether the layer dips upstream or downstream.

There are two exceptions to the preceding rule. The first is that a V-shaped pattern formed where a stream cuts across horizontal beds always points upstream. Thus,

73

FIGURE 8–1 Schematic cross sections of flat-lying strata (top), a monocline (middle), and a homocline (bottom).

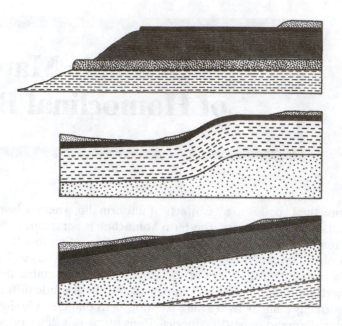

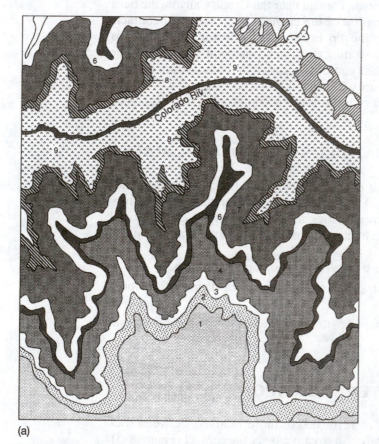

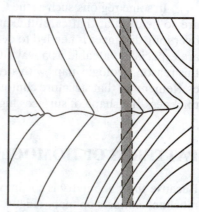

(b) Vertical bed.

(a)

FIGURE 8–2 **(a) Sketch map based on the geologic map of the Grand Canyon. Contacts of the rock units are parallel to topographic contours. (b) The contacts of a vertical bed or dike cut across the topography in straight lines. Note how the contacts show no deflection as they cross contours.**

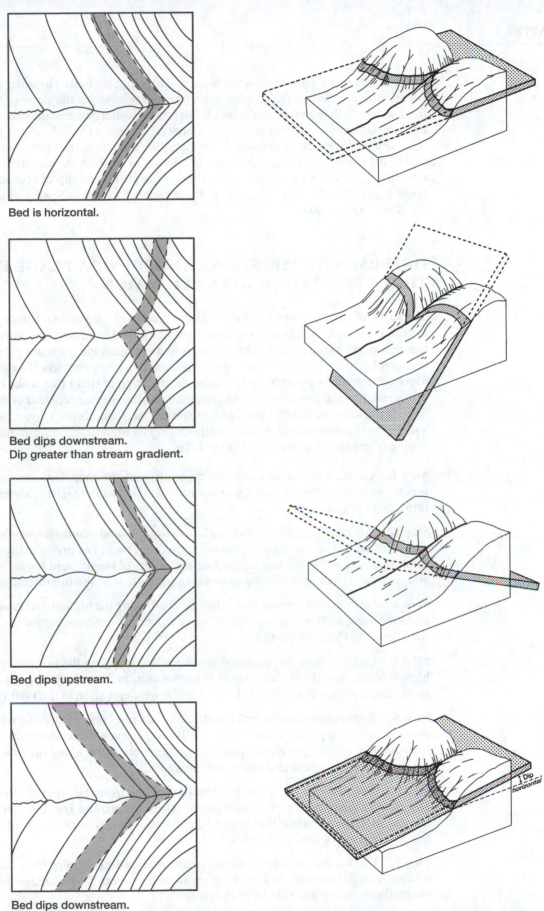

Bed is horizontal.

Bed dips downstream.
Dip greater than stream gradient.

Bed dips upstream.

Bed dips downstream.
Dip less than stream gradient.

FIGURE 8–3 **Block diagram and corresponding geologic maps illustrating the relationships between the dip of layers and the V-shaped patterns of contacts on geologic maps. (a) Horizontal bed. (b) Bed dipping downstream. (c) Bed dipping upstream. (d) Bed dipping downstream at an angle less than the slope of the stream.**

the presence of a V pattern does not always signify dipping beds. The other exception occurs where a contact dips downstream in a valley in which the slope of the valley floor is steeper than the dip of the bed. Under this unusual circumstance, the V-shaped contact pattern points upstream, in the opposite direction from the dip of the bed.

If a geologic map is drawn on a topographic base map, the approximate direction of dip can be determined by noting the direction in which the elevation along the contact decreases. An exact determination of the strike and dip of contacts can be made using elevation data on a bed at three points (see the following discussion of the three-point problem).

DETERMINING THE STRIKE AND DIP OF A PLANE FROM THREE POINTS OF KNOWN ELEVATION

The strike of a plane contact is the compass direction of a horizontal line on that contact. Thus, the strike of a planar contact can be determined from any two points on a contact that occur at the same elevation. Because many geologic maps are published on topographic bases, it is easy to determine the elevation along contacts. If structure contours are present on a map, they also show the direction of strike (the strike of the bed at any point on a structure contour map is parallel to the structure contour at that point).

If the location and elevation of three points on a contact are known, both strike and dip can be determined. Several methods may be used to accomplish this. The following is one such method (see Figure 8–4).

Step 1.   Connect the three points on an overlay of the map. Unless two of these points are at the same elevation, you can refer to them as the highest, lowest, and intermediate points.

Step 2.   Write the elevation beside each of the points, and determine which point occurs at an elevation intermediate between the other two. (The problem then is to determine where, along the line connecting the points of highest and lowest elevation, you can find a point that has the same elevation as that of the intermediate point.)

Step 3.   Draw a cross section along the line connecting the highest and lowest points using the same scale as that used for the map. This cross section will show the line from the highest to the lowest point.

Step 4.   Locate where the intermediate elevation occurs on the line connecting the highest and lowest point elevations in the cross section. Then project that point back to the line between the points of highest and lowest elevation in the map view.

Step 5.   Remember that the strike is the compass direction of any line on the map that connects two points of the same elevation. One such line connects the point of intermediate elevation with the point of the same elevation along the line connecting the points of highest and lowest elevation.

Step 6.   To determine the amount of dip, draw a line from the point of intermediate elevation in the direction of dip (perpendicular to the strike). The direction of dip is at right angles to the strike line, and the dip direction is on the same side of the strike line as the point of lowest elevation.

Step 7.   Draw a line parallel to the strike line and passing through the point of lowest elevation. If two of the points are at the same elevation, the bearing of the line between those points is the strike of the plane.

Step 8.   Construct a cross section along the line showing the direction of dip. You now have two elevations on the contact—the intermediate point and the lowest point. Connect these, and measure the angle of dip with a protractor.

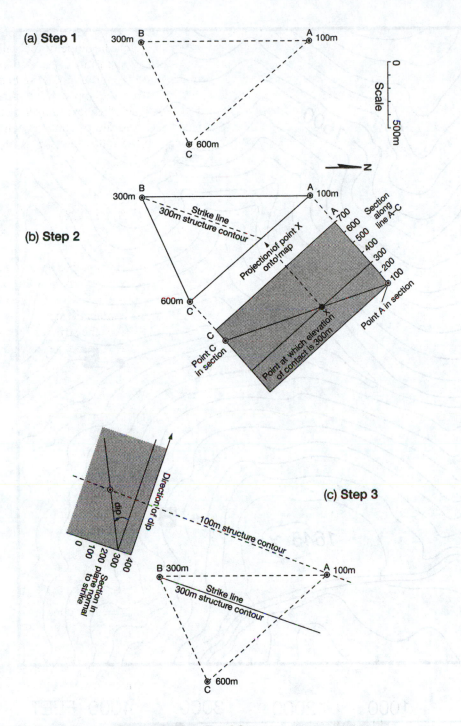

FIGURE 8–4   Steps in the solution of a three-point problem. See the text for discussion.

(a) **Step 1**

(b) **Step 2**

(c) **Step 3**

EXERCISE 8–1   Three-Point Problem

Refer to Figure 8–5.

1. What are the strike and dip of a plane contact that passes through points A, B, and C on the map in Figure 8–5a? Note that the right and left margins of the map are north–south lines.

   A. Strike _____ B. Angle of dip _____ Dip direction _____

2. What are the strike and dip of a plane contact that passes through points A, B, and D on the same map?

   A. Strike _____ B. Angle of dip _____ Dip direction _____

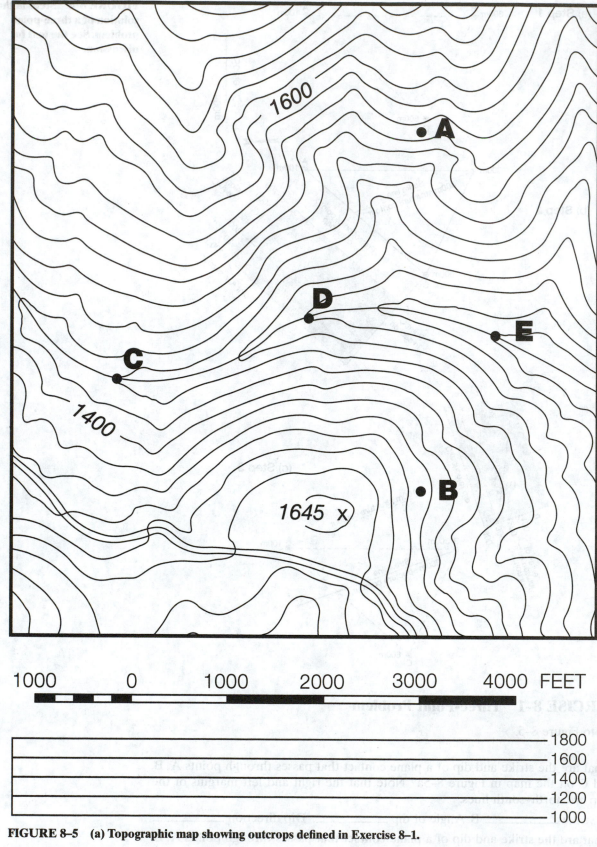

1000   0   1000   2000   3000   4000 FEET

1800
1600
1400
1200
1000

FIGURE 8–5   (a) Topographic map showing outcrops defined in Exercise 8–1.

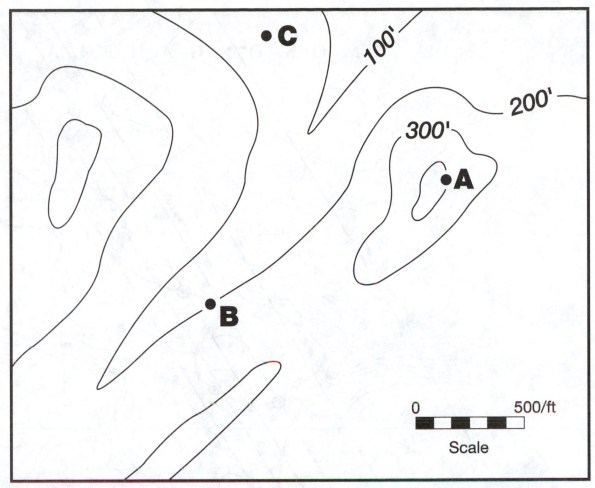

FIGURE 8–5 (b) Topographic map for use with Exercise 8–1.

3. What are the strike and dip of a plane contact that passes through points A, B, and E on the same map?

A. Strike _____ B. Angle of dip _____ Dip direction _____

4. Determine the strike, dip, and dip direction of the plane that crops out at the three points indicated in Figure 8–5b.

A. Strike _____ B. Angle of dip _____ Dip direction _____

TRACING PLANE CONTACTS THROUGH THE TOPOGRAPHY

The intersection of the contact between two rock units and the ground surface commonly forms complex map patterns. You have considered in a general way the types of V-shaped patterns that result where dipping layers cross valleys. If the strike, dip, and location of at least one outcrop of the contact of a plane layer are known, it is possible to trace that contact on a topographic map.

In the method outlined subsequently, a cross section of the dipping layer is drawn through the point where the contact crops out. Lines of equal elevation (structure contours) on the dipping contact are projected across the map. The contact should crop out on the ground where these lines intersect the ground surface. Follow the steps listed here (refer to Figure 8–6).

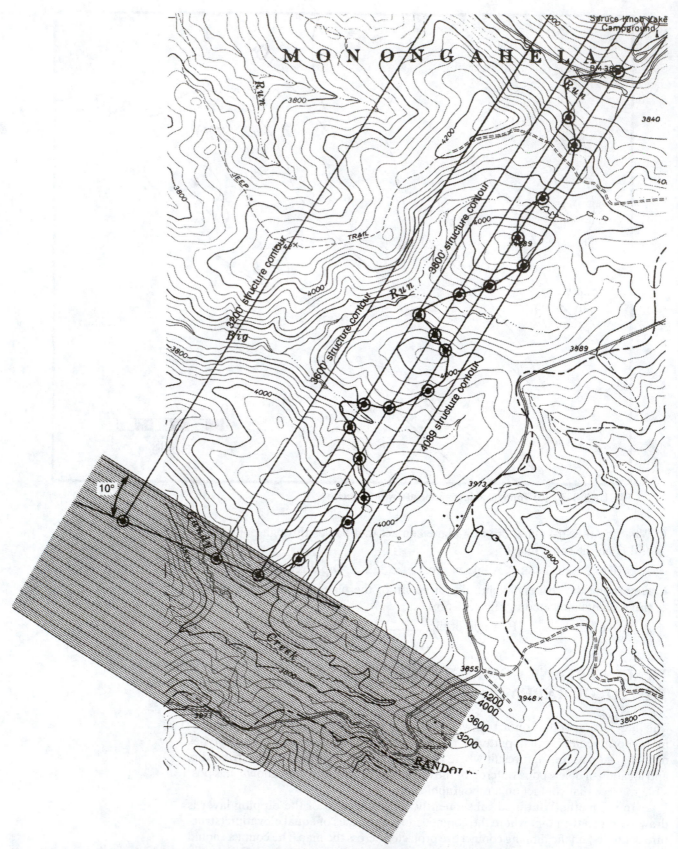

FIGURE 8–6   The method used to trace a contact through the topography. See the text for discussion.

Step 1.   Place a piece of graph paper on which the scale of the paper matches the scale of the map on the topographic map. It must be oriented so the strike of the contact to be traced is perpendicular to the cross section.

Step 2.   Project the point where the outcrop occurs onto the cross-section paper. Indicate this point on the cross-section paper at the same elevation at which it occurs on the topographic map.

Step 3.   Using the known dip of the layer, draw a cross section showing the bed.

Step 4.   Draw structure contours for a number of elevations on the layer across the map. (Sketch them in lightly.) For a plane surface, all structure contours will be parallel, straight lines.

Step 5.   Place a point on the map at each place where a structure contour on the contact crosses a topographic contour of the same elevation as the structure contour.

Step 6.   Connect the points with a smooth line. This line shows the trace of the bed across the topography.

Note: This same procedure can be used to trace a fault or even a folded bed, provided the fold is not plunging (e.g., the fold axis is horizontal) and its shape is the same in all cross sections across the fold.

EXERCISE 8–2   Tracing Contacts through the Topography

Refer to Figure 8–7.

A contact at the top of a rock unit that is 500 feet thick crops out along the road where it crosses the 1,600-foot topographic contour. This contact strikes north–south and dips 15°W. By using structure contours drawn on the top and bottom of this layer, trace the contacts across the map. Shade in the area where the rock unit should be exposed.

LAYER THICKNESS AND WIDTH ON MAPS

The thickness of a layer (strata, bed, etc.) is the distance between the top and bottom surfaces of that layer measured perpendicular to those surfaces. If vertical or horizontal layers are depicted on a topographic base, it is easy to measure thickness of strata. The thickness of a vertical layer is equal to the width of the outcrop on the map (Figure 8–8). The thickness of a horizontal bed is the difference in elevation between the top and bottom of the layer.

   If beds are inclined, layer thickness must be determined graphically or calculated using trigonometric relations. The graphical procedure is simple. Using a vertical scale that is the same as the horizontal scale of the map, draw a profile of the ground along a line perpendicular to the strike of the layer. Mark the location and dip of the upper and lower contacts, and draw in the contacts. Then measure the layer thickness using the map scale. The calculation of thickness using trigonometry is easy if the ground is flat; it is slightly more complicated when the ground is sloping (Figure 8–9). The slope of the ground surface can be determined from a topographic map by drawing a profile of the ground surface (see page 18) using the same vertical and horizontal scales.

   The following calculations may be used to determine layer thickness if the ground distances between the top and bottom of a layer are measured perpendicular to strike.

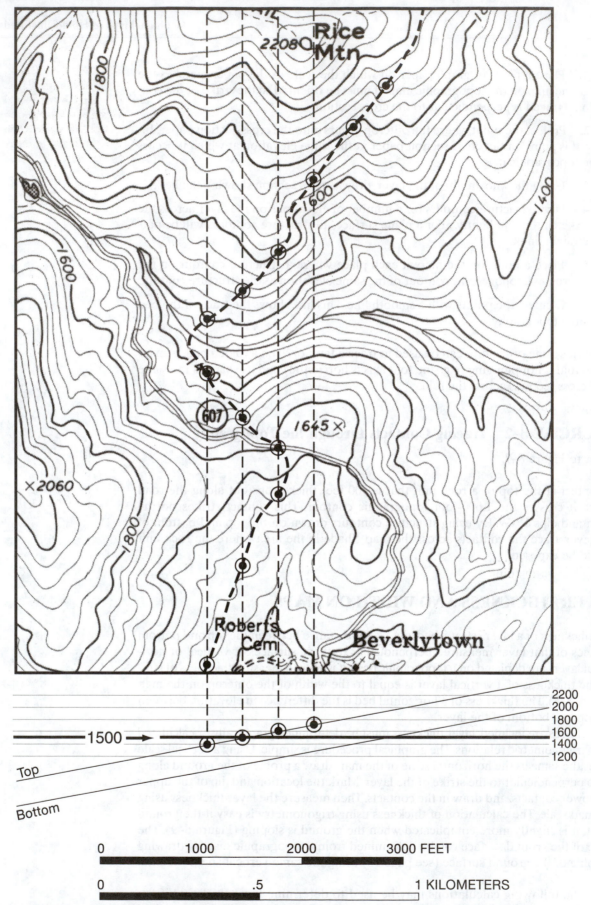

FIGURE 8–7 **Tracing contacts through the topography. A contact at the top of a layer that is 500 feet thick crops out along the road where it crosses the 1,600-foot topographic contour. This contact strikes north–south and dips 15° west. Using structure contours drawn on the top and bottom of this layer, these contacts are traced across the map. Structure contours for 1,400, 1,500, 1,600, and dashed lines show 1,700 feet on the map.**

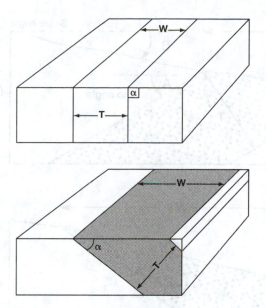

FIGURE 8–8   Block
diagrams showing the
relationship between the angle
of dip of a layer and the width
of the outcrop belt (measured
perpendicular to the strike of
the layers).

1. If the dip of the layer (A) and ground slope (B) are in opposite directions and the
   sum of the angles is greater than 90°:

2. If the layer dip (A) and ground slope (B) are in opposite directions and the sum
   of the angles is less than 90°:

3. If the layer dip (A) and ground slope (B) are in the same direction and the slope
   is less than the dip:

4. If the layer dip (A) and ground slope (B) are in the same direction, and if the dip
   angle is less than the slope:

EXERCISE 8–3   Layer Thickness and Width

Refer to Figure 8–10.

Determine the bed thickness in each of the following cases:

The scale in Figure 8–10 applies to all four maps.

1. Thickness of A: _____
2. Thickness of B: _____
3. Thickness of C: _____
4. Thickness of D: _____

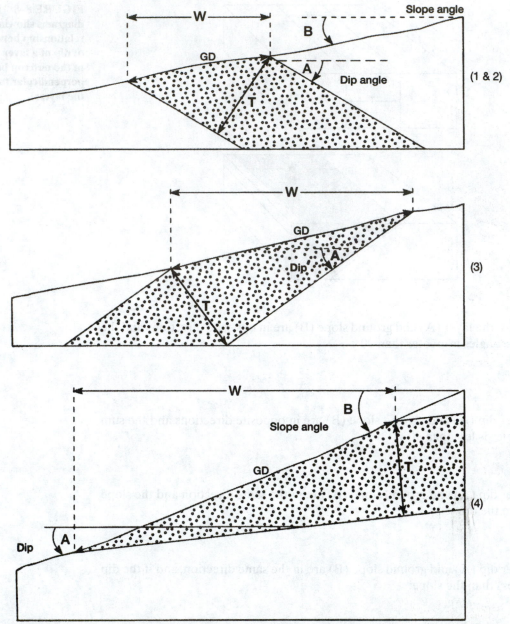

FIGURE 8–9 Cross sections illustrating the geometrical relationships involved in calculating layer thickness for various combinations of ground slope, layer dip, and dip direction. In all cases, the cross section is drawn perpendicular to the strike of the layer.

CONSTRUCTING CROSS SECTIONS OF HOMOCLINAL BEDS

A cross section depicts rock units as they would appear in a vertical slice through the earth's crust. In a few places, such as steep canal banks and in quarry walls, it is possible to actually see what a section looks like. Generally, geologists must construct cross sections on the basis of limited amounts of information. That information may be available in the form of drill holes from which samples or logs of various types have been taken. In some places, subsurface data are available from seismic surveys or

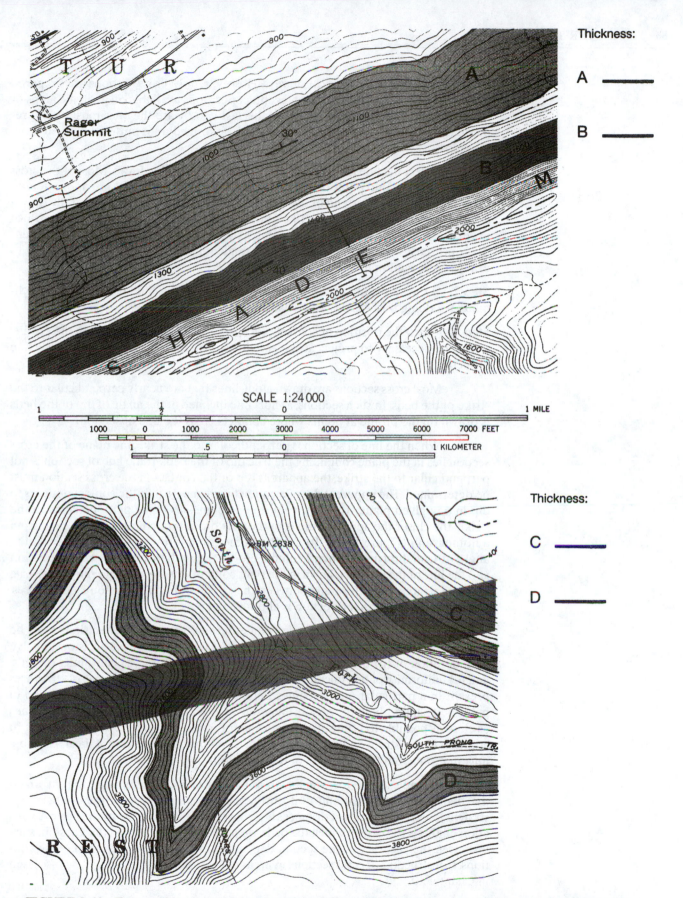

Thickness:

A ———

B ———

SCALE 1:24 000

Thickness:

C ———

D ———

FIGURE 8–10 Topographic maps showing the outcrop of layers. Note that the contacts of layer A dip 30° downslope. The contacts of layer B dip into the hillside. Contacts of layer C are straight as they cut across the topography. Contacts of layer D are parallel to the contours.

85

from other types of geophysical surveys. However, most cross sections are prepared using strike and dip information and contact positions available on geologic maps or taken from traverses. Cross sections of this last variety are the ones most often prepared by individuals making geologic maps. The procedure is outlined in the following steps:

Step 1. Select the line along which the cross section is to be drawn. Most of the cross sections in books and published papers are drawn along lines that are approximately perpendicular to the trend of the structure. Thus, section lines are generally drawn normal (perpendicular) to the strike of the beds. Most cross sections are also drawn in vertical planes. In areas of highly complex geology, it may not be possible to identify a primary structural trend on the geologic map. Nevertheless, it is possible to draw a cross section in any desired direction.

Step 2. Draw a profile of the ground surface along the line of section. The technique for doing this was described earlier on page 22.

Step 3. Mark the position of contacts and strike and dip information on the profile. It may be necessary to project strike and dip information from the line into the line of section.

Most cross sections are drawn along lines that are nearly perpendicular to the strike of the beds. In such sections, a short line inclined at the angle of dip of the beds is drawn on the profile at points where contacts cross the line of section.

When the line of section is perpendicular to the strike, the plane of the cross section lies in the plane containing the true dip of the beds. If the line of section is not perpendicular to the strike, the apparent dip of the contact in the cross section must be determined. The **apparent dip** of a contact is the angle that the contact appears to dip in any given plane. The closer that plane comes to being perpendicular to the strike, the closer the apparent dip is to the true dip. Apparent dips can be read from a nomograph (Figure 8–11). To use the nomograph, draw a straight line from the angle of true dip (the left-hand column) to the angle between the strike direction and the line of the section (shown in the right column). In the example shown, the true dip is 43 degrees, and the angle between the strike and line of section is 35 degrees. The apparent dip in that direction is about 28 degrees.

Step 4. Once all surface data have been entered on the profile, decisions must be made regarding how to project that data into the subsurface. These decisions involve consideration of (a) the probable geometry of the structure, (b) continuity of beds, and (c) uniformity of bed thickness. At this point, knowledge of the types of structures present in a particular area becomes important. This information may be found in published papers about the region. Generally, a cross section is redrawn until it comes as close as possible to fitting all the data and information you have about the area. In addition to the strike and dip information, it is important to have the best available information about the thickness of the rock units.

The simplest case for which a cross section may be prepared is that in which the beds are homoclinal and uniform in thickness. In that case, the contacts are straight lines spaced apart at distances appropriate to the thickness of the beds. A variation on this is the case in which the beds are homoclinal, but vary in thickness. In such instances, the accuracy of the section depends on the knowledge of thickness variations. Construction of cross sections in areas where folds are present will be discussed in Chapter 10.

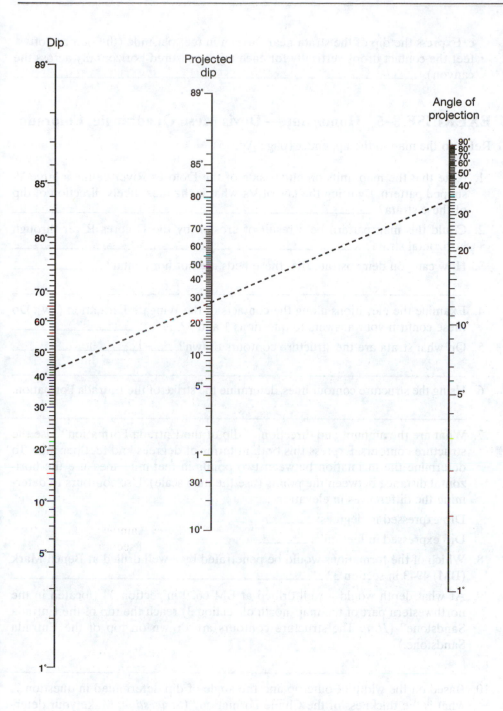

FIGURE 8–11 A nomograph used to convert true dips to apparent dips (or vice versa). (From H. S. Palmer, 1919, New graphic method for determining the depth and thickness of strata and the projection of dip: in "Shorter Contributions to Geology—1918," United States Geological Survey Professional Paper 120-G.)

EXERCISE 8–4   Homoclines—Grand Canyon, Arizona

Refer to the map in the appendix, page Ap2–13.

1. Based on the relationship between contacts and contours, are the strata near the top of the canyon horizontal? _____

2. Determine the elevation of the edge of the canyon rim on the north and south side of the canyon. The edge of Kaibab Limestone (Pk) forms the rim of the canyon on both sides.

   a. Elevation of the north rim: _____

   b. Elevation of the south rim: _____

c. Express the dip of the strata near the rim in feet per mile (the number of feet the contact drops vertically for each mile measured horizontally across the canyon): _____

EXERCISE 8-5   Homoclines—Davis Mesa Quadrangle, Colorado

Refer to the map in the appendix (page Ap2–5).

1. Note that the map units on either side of the Dolores River define a large V-shaped pattern. By using the law of Vs, what is the most likely direction of dip of these strata?_____

2. Could this map pattern be a result of erosion by the Dolores River through horizontal strata? _____

3. How can you demonstrate that these beds are not horizontal? _____

_____

4. Examine the elevations along the contacts of the Wingate Formation (Jw). Do these confirm your answers to questions 1 and 3?_____

5. On what strata are the structure contours drawn? _____

_____

6. Using the structure contour lines, determine the strike of the Entrada Formation.

_____

7. What are the amount and direction of dip of the Entrada Formation? Use the structure contours. Express this both in terms of degrees and feet per mile. To determine the inclination between two points in feet/mile, measure the horizontal distance between the points (use the map scale). Use contours to determine the differences in elevation.

   Dip expressed in degrees: _____

   Dip expressed in feet/mile: _____

8. Which of the formations would be penetrated by a well drilled at Bench Mark (BM) 4943 in section 3? _____

9. At what depth would a well drilled at BM 6629 in section 34 (located in the northwestern part of the map, north of section 3) reach the top of the Entrada Sandstone? (*Hint:* The structure contours are drawn on top of the Entrada Sandstone.)

_____

10. Based on the width of outcrop and the angle of dip determined in question 7, what is the thickness of the Chinle Formation? (*Suggestion:* Make your determination near the bottom edge of the map in section 10, where you can measure the width of outcrop perpendicular to the strike of Chinle; remember that the width must be measured perpendicular to the strike of the unit.)

_____

EXERCISE 8–6   Homoclines—Ouray Quadrangle, Colorado

Refer to the map in the appendix, page Ap2–23.

1. Describe the structure of the Mesozoic sedimentary units in this map area.

   _____

2. Determine the elevation of the base of the Mancos Shale on the east and west side of the valley in which Ouray is located. In which direction does the contact dip? What is the approximate change in elevation (express in feet per mile)?

   _____

   _____

   _____

3. Study the mineralized veins (red lines). Identify places where the mineralization has taken place along faults. Why might mineralization occur along a fault zone?

   _____

   _____

   _____

EXERCISE 8–7   Cross Sections of Homoclinal Beds

Refer to Figure 8–12.

Draw a cross section along the line A–B in the space provided in Figure 8–12. A profile of the land surface is shown along the section line. First, draw the cross section using the dip information provided along the line without correcting dip information for apparent dip. Then redraw the section using apparent dips obtained from Figure 8–11. Note that the dip of beds decreases from right to left. This indicates that the beds are either wedge-shaped or slightly folded.

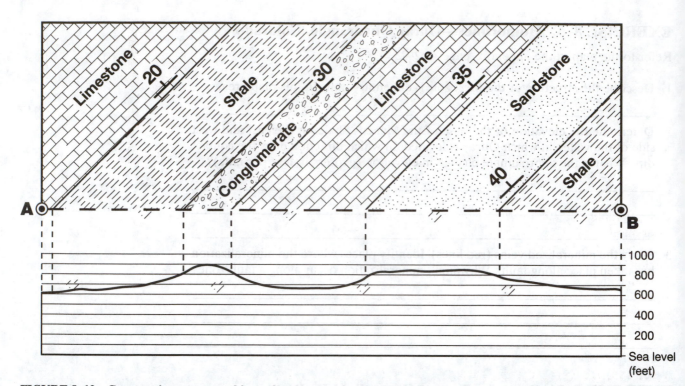

FIGURE 8–12   Constructing a topographic profile along section line is the first step in preparing all cross sections. Next, strike and dip data from the geologic map are projected into the section line. If the dips are low, or if the angle between the line of the cross section and the strike of the beds is great, true dips should be converted to apparent dips in the cross section. Use the nomograph in Figure 8–11 to make this conversion. Third, the dips of contacts are drawn on the topographic profile. Finally, if the cross section is to be drawn freehand, contacts are projected beneath the ground profile and connected by smooth, curved lines. Unless other information is available, the thickness of the rock units should be maintained constant across the section.

CHAPTER 9

Unconformities

A stratigraphic sequence records a history of sedimentation. In some places, one layer after another can be seen deposited in an orderly succession, forming a continuous record of sedimentation (Figure 9–1). In other places, the sequence is broken in what geologists call **unconformities.** They come into existence as a result of interruptions in sedimentation. Generally, they are also erosion surfaces (the former ground surface), now exposed only where the stratigraphic succession containing the missing interval is exposed at the present ground surface.

UNCONFORMITY PATTERNS ON GEOLOGIC MAPS

Three distinctly different types of unconformities are illustrated in Figures 9–2 and 9–3. Each is characterized by a distinct and different map pattern on geologic maps.

1. Angular unconformities are breaks in the stratigraphic record, separating rock units that have different structural attitudes above and below the break. By definition, an angular discordance exists between the units on either side of an angular unconformity. On a geologic map, angular unconformities are lines of contact across which strata have different strikes and dips. Unless the entire section has been deformed after the unconformity formed, strata above an angular unconformity have lower dips than those below the unconformity. A prominent angular unconformity separates the folded and faulted Paleozoic strata in the Appalachian and Ouachita Mountains from the relatively flat-lying Mesozoic and Cenozoic strata of the Atlantic and Gulf coastal plains. The way this unconformity appears on the Arkansas State map (appendix, page Ap2–3) is representative of the structural relations commonly seen across angular conformities.

2. Nonconformities are erosion surfaces separating crystalline rocks, usually igneous intrusions or metamorphic complexes, below the surface from sedimentary rocks above it. Examples of nonconformities exist along the contact between the Precambrian rocks of the Canadian Shield, the Ozark Dome, the Adirondack Dome, the Black Hill Uplift, and the surrounding sedimentary cover in each of these areas.

3. Disconformities are breaks in stratigraphic successions marking erosion or nondeposition in an otherwise conformable sequence. The disconformity surface is generally irregular; if no irregularities occur on the surface, it is called a **paraconformity.** On a geologic map, disconformities and paraconformities are identified by the absence of a rock unit that is present in other places or by the thinning out of a unit.

FIGURE 9–1   Block diagram showing three layers with conformable contacts.

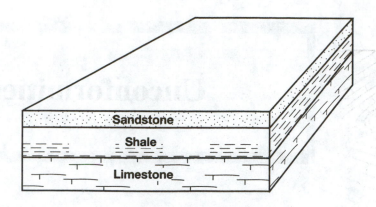

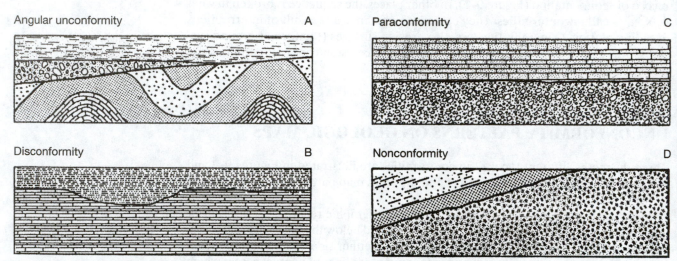

FIGURE 9–2   Schematic cross-sectional representations of various types of unconformities. Beds are parallel to one another above and below a disconformity, but some evidence of topographic relief is present. Relief features are not evident at paraconformities.

EXERCISE 9–1   Unconformities—Pittsfield East Quadrangle, Massachusetts

Refer to the map in the appendix, page Ap2–21.

1. Describe the contact between the units of marble (Ose and Osd) and the underlying Csc.

2. Explain how this contact may have formed.

(a)

(b)

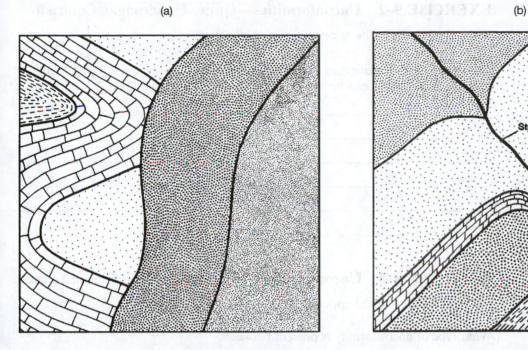
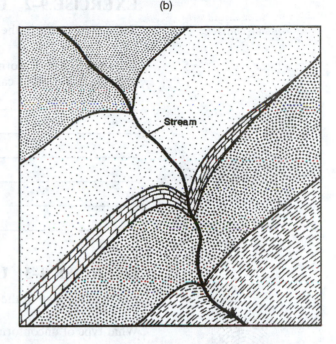

(c)

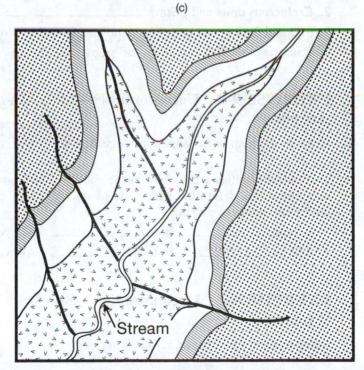

FIGURE 9–3 (a) Schematic map of an angular unconformity. Beds below the unconformity are shown folded, but any angular difference between layers above and below the unconformity is sufficient to warrant calling it an angular unconformity. (b) Schematic map of a disconformity. Note how the limestone bed marked with the brick pattern thins and finally disappears leaving what appears to be a conformable sequence. (c) Schematic map showing a nonconformity overlain by flat-lying beds.

EXERCISE 9–2   Unconformities—Ouray Quadrangle, Colorado

Refer to the map in the appendix, page Ap2–23.

A number of unconformities are present in this region (see the legend). Which of these unconformities can be identified on the map, and where are they exposed?

_____

_____

_____

_____

_____

EXERCISE 9–3   Unconformities—Brownwood and Llano, Texas

Refer to the map in the appendix, page Ap2–7.

What type of unconformity is present between:

1. Cretaceous unit (Kh = Ka) and the Cambrian rock units? _____
2. Cretaceous units and pЄtm? _____
3. MD and Ordovician units? (near northern border of map) _____
4. Cambrian unit Єrh and pЄtm and pЄps? _____

EXERCISE 9–4   Unconformities—Mule Mountain Quadrangle, Arizona

Refer to the map in the appendix, page Ap2–21.

Two unconformities are present in the southwestern part of this map. Identify the unconformities, determine what type of unconformity each is, and indicate what lies below and above each.

a. Type: _____   Below: _____   Above: _____

b. Type: _____   Below: _____   Above: _____

EXERCISE 9–5   Unconformities—Bull Lake West Quadrangle, Wyoming

Refer to Figure 9–4. Scale: 1:24,000

Stratigraphic Units (modified after the original map):

Qal	Alluvium and colluvium	Quaternary
Qp	Pinedale glacial till and outwash gravel	
Qb	Bull Lake glacial till and outwash gravel	
Qbl	Bull Lake interglacial soil	
Qw	Washakie Point glacial till	

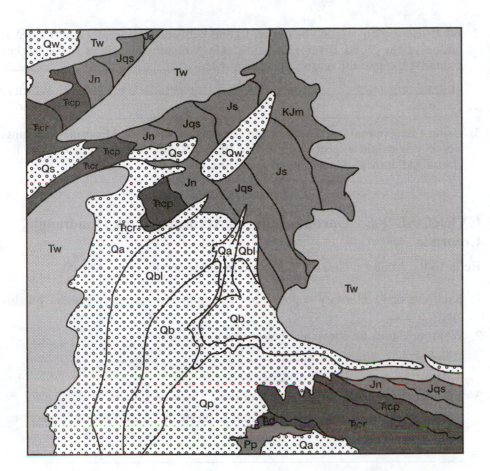

FIGURE 9–4 Simplified portion of the geologic map of the Bull Lake Quadrangle. (After J. Murphy and G. Richmond, 1965. United States Geological Survey Map GQ-432.)

Tw	Wind River Formation	Tertiary
KJm	Cloverly and Morrison Formations	Cretaceous and Jurassic
Js	Sundance Formation	Jurassic
Jgs	Gypsum Spring Formation	
Jn	Nugget Sandstone	
Ŧcp	Popo Agie Member of Chugwater Formation	Triassic
Ŧcr	Read Peak Member of Chugwater Formation	
Ŧd	Dinwoody Formation	
Ppc	Park City Formation	Permian

Note: The older consolidated bedrock formations (Cretaceous to Permian age) are all sedimentary rocks that have been tilted along the flank of the Wind River Range, a large uplifted block of Precambrian gneisses and its cover. The Tertiary Wind River Formation is composed of semiconsolidated gravels derived from the Wind River Range after its uplift. Still younger glacial deposits and recent alluvium and colluvium have been deposited during Pleistocene and Recent time.

1. What is the stratigraphic relationship between the Wind River Formation and the Mesozoic units? (Is the sequence a normal stratigraphic sequence? Are unconformities present in this area? If so, what kind?) _____

_____

_____

2. What relationship would you expect to find between topography and the distribution of the Wind River Formation? (Is it mainly in valleys or on higher ground?) Explain the reason for your answer. _____

_____

3. Using a transparent overlay, draw the contacts between the Mesozoic rock units where you would expect to find them if the Wind River Formation and the glacial and alluvial deposits were stripped away.

EXERCISE 9–6   Unconformities—Atkinson Creek Quadrangle, Colorado

Refer to the Atkinson Creek Quadrangle in the appendix (page Ap2–5).

1. Exclusive of Quaternary deposits, is an angular unconformity present on this map?_____

2. What rock units lie above and below this unconformity?

   a. Below: _____

   b. Above: _____

3. Examine the outcrop width of each of the Jurassic rock units.

   a. What evidence, if any, do you see that suggests that a disconformity may be present?

   _____

   b. Which rock unit is missing at the disconformity? _____

(*Hint:* Refer to the stratigraphic column and compare it with the sequence of rock units shown on the map.)

EXERCISE 9–7   Unconformities—Grand Canyon, Arizona

Refer to the map in the appendix, page Ap2–13.

1. Using a piece of tracing paper, trace the contact of the Great Unconformity (the unconformity that separates the Precambrian-age rocks from Cambrian-age rocks).

2. What type of unconformity is the Great Unconformity? What rock units are above and below the unconformity? _____

_____

_____

_____

_____

3. What type of unconformity is the unconformity between the Younger and Older Precambrian rock units? _____

4. Rock units of what age are missing in the canyon? _____

EXERCISE 9–8   Unconformities—Arkansas State Geologic Map

Refer to the map in the appendix, page Ap2–3.

A major angular unconformity runs diagonally across this map.

1. What age strata lie immediately below and above the unconformity in the southwestern corner of the map?

   a. Below: _____

   b. Above: _____

2. What age strata lie immediately below and above the unconformity near Little Rock?

   a. Below: _____

   b. Above: _____

3. How many different rock units lie immediately above the unconformity on this map?

   _____

4. How can you explain the observation that rock units of different ages and different formation lie above this unconformity at different places along the unconformity?

   _____

5. What types of structural features (e.g., folds, faults, homoclines, etc.) lie beneath the unconformity?

   _____

CHAPTER 10

Folds on Geologic Maps

FOLD GEOMETRY

Depending on the conditions under which folding took place, fold geometry may range from simple curvilinear folds such as the Waterpocket fold illustrated in Figure 10–1a to exceedingly complex shapes such as those most often seen in metamorphic rocks (Figure 10–1b). Less complex, geometrically simple folds of the types most often found on geological maps of sedimentary rocks are used in this discussion. The terms commonly used to describe such folds are illustrated in Figure 10–2. The crest line of a fold is a line drawn along the highest part of a folded layer. The **trough line** similarly indicates the lowest part of the fold. Portions of the folded layer that lie between crests and troughs are called fold limbs. The plane that divides a fold into two halves that are more or less mirror images of one another is called the **axial plane** or **axial surface** of the fold. The line defined by the intersection of this plane with the surface of the ground is the **axial trace.** Some geologic maps contain axial traces of large folds. The **fold axis** is an imaginary line formed where the axial plane intersects a folded layer. The axis may be horizontal, vertical, or inclined. If it is inclined, the angle measured in a vertical plane between the axis and horizontal projection of the axis is called the plunge (Figure 10–2c).

Folds are described as upright, asymmetrical, or recumbent, depending on the orientation of the axial surface, also called the axial plane. Folds are described as symmetrical or asymmetrical, depending on the symmetry of the halves on either side of the axial surface, when viewed in cross section along the fold axis (Figure 10–3).

FOLD PATTERNS ON GEOLOGIC MAPS

The pattern formed by the contacts of folded rock units on a geologic map depends on such factors as the size of the fold, the choice of map units, topographic effects, and the geometry of the folded rocks.

1. The size of the fold relative to the area covered by the map.   Many maps show only portions of the limb of a large fold. The scale of the map and the thickness of map units affect the appearance of fold patterns. For example, the folds shown on state or national maps are generally so large that the topography does not have a great effect on the map pattern. Examine the folds shown on the Arkansas State map (appendix, page Ap2–3). Map units must be very thick and the folds must be large to show on the map at this scale.
2. The selection of map units.   Contacts of large folds may appear to be homoclinal if the map covers only a portion of the fold. Small folds within formations rarely show up on maps. Generally, folds are evident on geologic maps only if contacts of map units are affected by the folding.

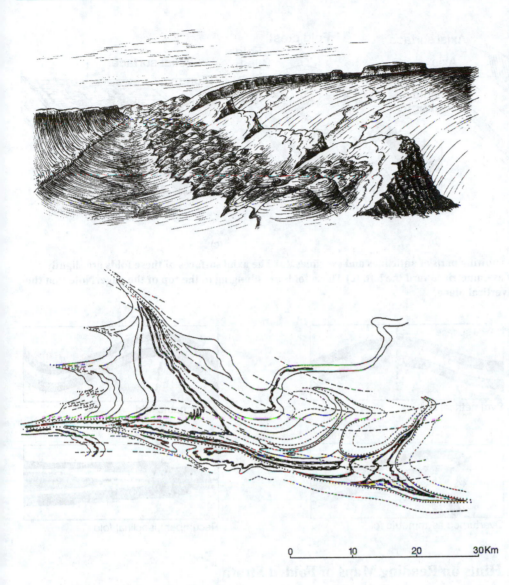

FIGURE 10–1   **(a) Sketch of part of the Waterpocket fold (a monocline) in Utah. This is an example of a geometrically simple structural feature. (After G. K. Gilbert.) (b) Geologic map of part of the Gwanda greenstone belt in the Rhodesian craton. The rocks here have been highly deformed. Early formed folds were refolded producing a highly complex geometric pattern. (From "Northern margin of the Limpopo mobil belt, Southern Africa,"** *Geological Society of America Bulletin,* **Coward, M. P., James, P. R., and L. Wright. Reproduced with permission of the publisher, the Geologic Society of America, Boulder, Colorado, USA. Copyright © 1976 Geological Society of America.)**

0   10   20   30Km

3. Topographic effects.   On maps at scales of 1:24,000 or less, even perfectly planar contacts may appear as complex patterns as a result of the way the contact between map units intersect the ground surface. Similar effects determine map patterns of folded rocks.

4. The geometry of the folded strata, including the symmetry of the fold, the attitude of the limbs, the curvature of the fold, and the plunge of the fold axis.   Fold shapes range from simple, symmetrical anticlines and synclines to complex refolded folds of intricate form.

Because the appearance of a fold on a geologic map depends on the shape of the topography as well as on the geometry of the fold, maps of real folds may appear quite different from schematic illustrations, which do not show effects of topographic relief (Figure 10–4). Actual geologic map patterns may closely resemble the schematic patterns of illustrations if the mapped area has subdued relief, or if the size of the structural feature is large relative to the amount of relief. For this reason, large folds on state and national maps often resemble schematic maps more than do folds on 7.5-minute quadrangle maps.

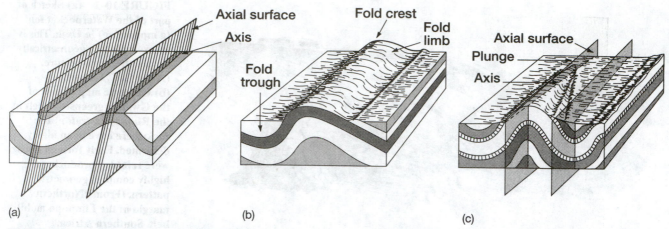

FIGURE 10–2   Block diagrams showing parts of anticlines and synclines. (a) The axial surfaces of these folds are slightly tilted to the right. (b) This fold is asymmetric toward the left. (c) These folds are plunging to the top of the page. Note that the angle of plunge is measured in a vertical plane.

FIGURE 10–3   Examples of various types of fold symmetry as viewed in cross sections normal to the fold axis.

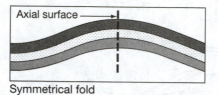

Symmetrical fold

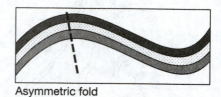

Asymmetric fold

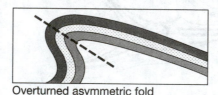

Overturned asymmetric fold

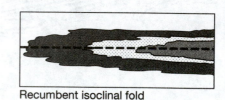

Recumbent isoclinal fold

Hints on Reading Maps of Folded Strata

1. When the folded strata are of different resistance to erosion, folds are likely to be reflected in the topography.

2. V-shaped patterns formed where streams cut contacts of folded layers reveal the direction of dip of those contacts just as they do for planar layers. Do not confuse the V-shaped patterns that form where streams cut across fold limbs with the V-shaped patterns formed where folds plunge (Figure 10–4).

3. If a layer is uniform in thickness, and if the area has low relief, the relative width of the outcrop at different places along a fold or on different limbs of a fold is a good indication of dip. Thus, if the outcrop width of one limb is much greater than that of the other, the dips of the two limbs are different, and the fold is probably asymmetric in cross section.

4. Folds that have parallel limbs are called isoclinal folds. The limbs of isoclinal folds dip in the same direction. Both anticlines and synclines may be isoclinal. If V-shaped patterns are present where streams cut across the limbs, the Vs on both limbs point in the same direction. Fold hinges of isoclinal folds are commonly sharp and narrow.

5. To obtain some idea of what the cross section of a structure should look like, it is useful to use a technique known as down-plunge viewing. If you know that a fold

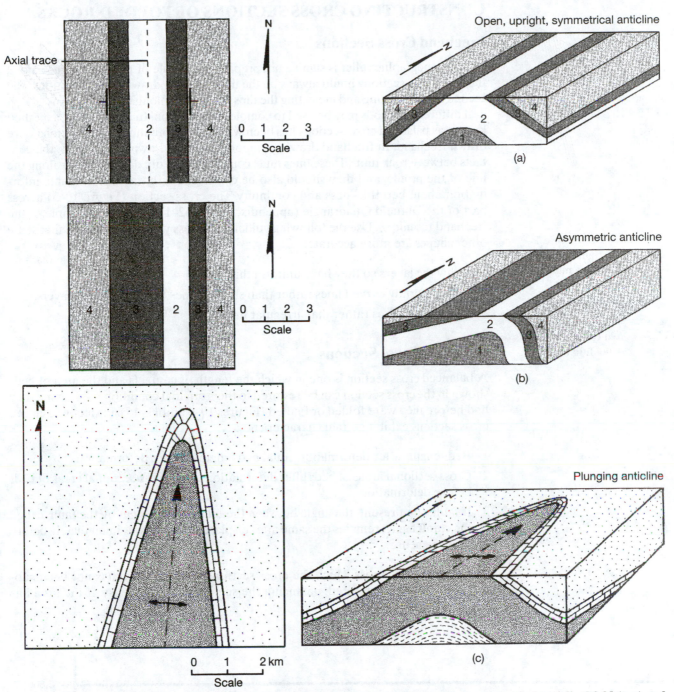

FIGURE 10–4   (a and b) Block diagrams and corresponding geologic maps for an open, symmetrical anticlinal fold (top) and an asymmetric anticlinal fold (bottom). Neither fold is plunging. (c) A plunging anticline with corresponding map (right).

is plunging at an angle of 30 degrees, place the map on a flat surface and rotate the map until you are looking at it in the direction of the plunge of the fold. Then raise or lower your head until your line of sight is inclined at the same angle as the angle of plunge. In this position, you should see approximately what a cross section drawn perpendicular to the fold axis looks like. The same technique can be used to look down-dip to see what faulted beds look like in section.

CONSTRUCTING CROSS SECTIONS OF FOLDED ROCKS

Freehand Cross Sections

Where topographic relief is significant, preparing a profile of the land surface along the line of the section should always be the initial step in drawing cross sections. The second step is locating and indicating the dips of contacts along the profile. Any of several different methods may be used to complete the section. In recent years, a method known as balanced cross sections has been widely used, but most older sections were drawn freehand. In freehand drawing, smooth lines are drawn to represent the contacts between rock units. These lines must conform to the dip data available along the line of the profile, and they should also be consistent with the best available information about bed thickness and continuity. The cross section (Figure 10–5) across part of the Duffield Quadrangle (appendix, page Ap2–11) is a good example of the freehand technique. Use the following guidelines unless you have evidence that some other shapes are more accurate:

1. Draw the layers so they have uniform thickness.
2. Use uniformly curved lines rather than straight lines to represent the layers.
3. Use smooth lines rather than irregular lines.

Balanced Cross Sections

A balanced cross section is one in which the length of contacts and the area of beds shown in the cross section can be restored to their original condition—the shape they had before they were folded or faulted (Figures 10–6 and 11–15). Perfectly balanced cross sections exhibit certain characteristics:

1. Bed lengths after deformation equal bed lengths before deformation.
2. Cross-sectional areas of beds after deformation equal cross-sectional areas of beds before deformation.
3. If faults are present, the angle between the fault and beds cut by the fault, called the **cutoff angle,** remains the same after displacement as it was immediately before displacement.

If the area across which the cross section is being drawn has not been subsequently folded or faulted, balancing the section should be easy. If the area has

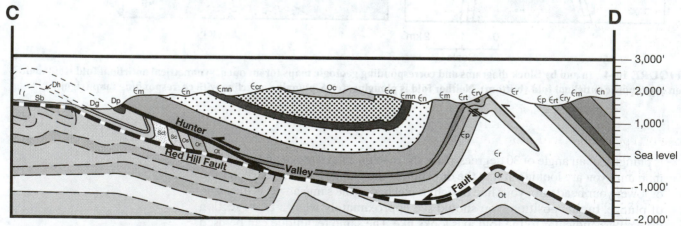

FIGURE 10–5   This cross section is drawn across the asymmetric syncline shown on the map of part of the Duffield Quadrangle (see the map in the appendix).

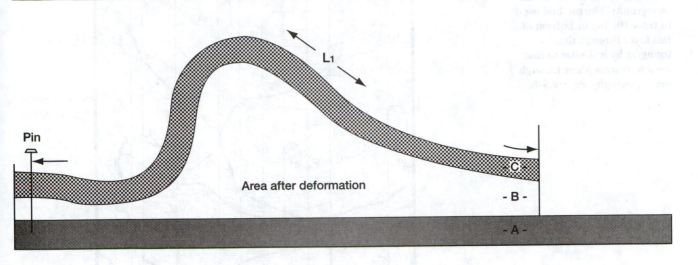

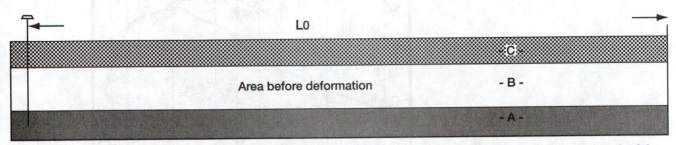

FIGURE 10–6   In a balanced cross section the original length of layers before deformation (L_0) is equal to the length of the same layers after deformation (L_1), and the cross-sectional area of each layer should be the same before and after deformation. Using the grid, check the top cross section to see if this is a balanced cross section.

been deformed, the shape of beds in the section will have changed from their original form, but neither volume nor bed lengths should have changed much during deformation.

To test for balance, measure the length of each contact shown on the section. They should be approximately the same lengths. With a planimeter or digitizing tablet, measure the area of each bed. It should be possible to reconstruct an undeformed section from the preceding data. (*Note:* The preceding test is not completely valid unless the cross section extends far enough across the structure to reach a point at which the amount of slip between beds is minimal.) See Woodward, Boyer, and Suppe (1990) for more detailed information. Elementary discussions of balanced sections with exercises are also available in Marshak and Mitra, 1988.

TRACING FOLDS THROUGH THE TOPOGRAPHY

The same technique used to trace the line of intersection of a plane bedding contact or a plane fault across the land surface (see page 79) may be used to trace the contacts of folded layers through the topography (Figure 10–7). This technique is a useful way of learning how folded layers may appear on topographic maps, but application of this technique to real folds is limited. The method gives reliable results only if the fold axis is horizontal and if the shape of the fold in cross section remains uniform. Most real folds change in cross-sectional shape and plunge at one or both ends.

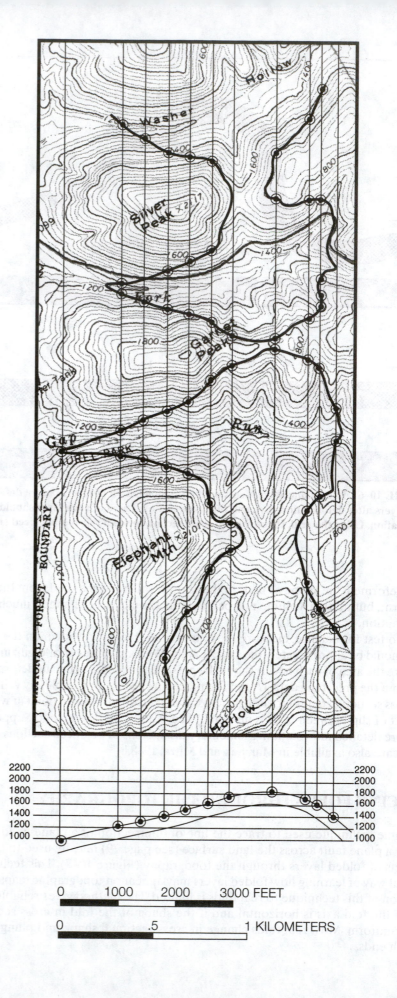

EXERCISE 10–1   Tracing Folds through the Topography

Refer to Figure 10–7.

The trace of the outcrop of the upper contact of a folded unit is already drawn on this map. The unit is approximately 200 feet thick, and is shown in the cross section. Locate the trace of the lower contact of this unit, and color the outcrop of the unit on the map.

STRUCTURE CONTOUR MAPS OF FOLDED STRATA

Structure contour maps provide one of the best methods of depicting the shape of folded strata. Structure contours are shown on the maps of Pine Mountain, Atkinson Creek, Davis Mesa, and Salem quadrangles in the appendix. Unfortunately, geologists rarely have access to sufficient subsurface data to draw accurate structure contour maps. Structure contours are generally drawn on the top of a distinctive rock unit that is readily identified in well logs or on seismic sections. The surface on which the contours are drawn is referred to as a contact or horizon. If the horizon on which the structure contours are drawn crops out, elevation data are available all along the contact, but portions of the horizon above ground level have been removed by erosion. If the horizon lies entirely in the subsurface, information used to construct the structure contours comes from well logs and/or seismic data. For this reason, most structure contour maps depict areas where oil and gas are present. The Oriskany Sandstone is one of the principal gas-producing horizons in the Appalachian Basin. The structure contour map of an area on the edge of this basin (Figure 10–8) illustrates folds of the Valley and Ridge province. Note that the spacing of contours on the fold limbs clearly shows the asymmetric shape of these folds, and the closed contour lines indicate that the folds plunge toward both the northeast and southwest.

EXERCISE 10–2   Folds—Northern and Western Flanks of the Black Hills, Wyoming

Refer to the sketch maps, Figures 10–9 and 10–10. Scale: 1:96,000

Small circles indicate gas wells with a cross through them.

1. The structure contours are drawn at a contour interval of 100 feet. The datum is sea level. Contours are drawn on top of the Fall River Sandstone, which does not outcrop in this map area. Using the structure contours, draw cross sections along the lines indicated by letters A, B, C, and D. Draw cross sections with 2× and 10× vertical exaggeration for each line. (*Note:* At the scale of this map, 0.1 inch equals approximately 800 feet.)

2. Using a transparent overlay, trace the contacts of the bedrock units where they are most likely to pass beneath the Quaternary alluvium. Draw the axial traces of the folds on the overlay.

3. Are these folds plunging? If yes, in what direction? _____

_____

4. Are the folds symmetrical or asymmetrical? _____

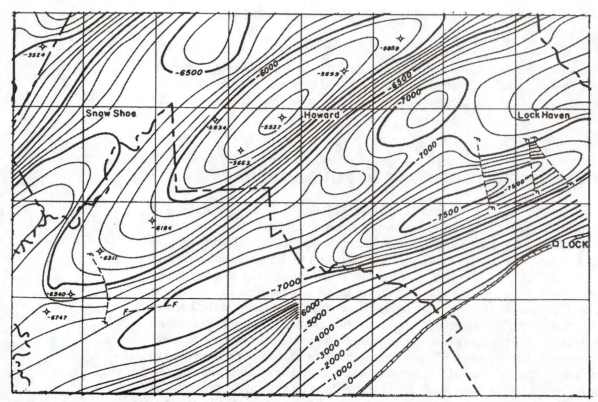

FIGURE 10–8   A structure contour map of an area in Pennsylvania. Contours are drawn on the top of a Devonian sandstone, the Oriskany Sandstone. (From Cate, A. S., et al., 1961, Subsurface structure of plateau region north-central and western Pennsylvania on top of Oriskany formation: Pennsylvania Geological Survey, 4th series map.)

5. If the folds are asymmetrical, which is the steep limb?_____

_____

6. Why are the structure contour lines smoother than the mapped contacts?

_____

7. Based on the shape of the contacts on this map, do you think the beds have low, moderate, or steep dips? The relief in this area is very low. _____

_____

_____

EXERCISE 10–3   Structure on the Arkansas State Geologic Map

Refer to the map in the appendix, page Ap2–3.

1. Using an overlay, sketch the area in which Tertiary rock units outcrop, omitting all Quaternary deposits that are shallow alluvial materials. The total thickness of Tertiary rocks is no more than a few thousand feet thick. How would you describe the structure of the Tertiary rock units shown on this map? Are they folded, faulted, or homoclinal?

_____

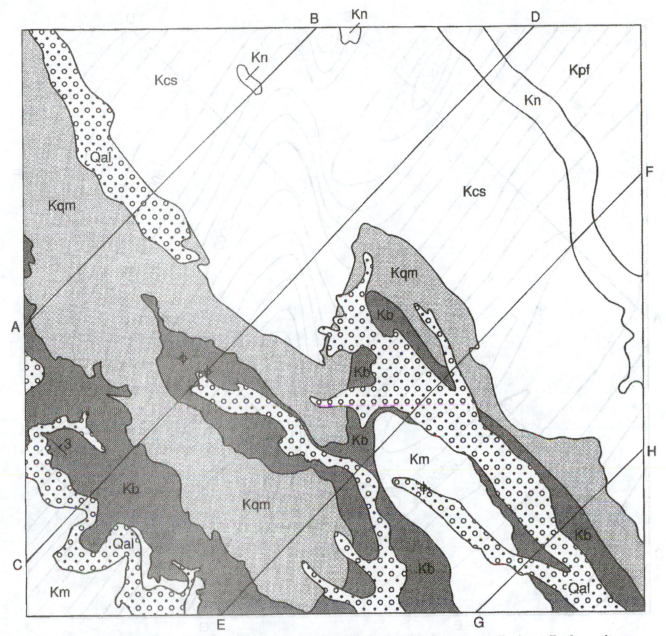

FIGURE 10–9   Geologic sketch map of an area north of the Black Hills. Qal is Quaternary alluvium; all other units are Cretaceous in age. Km is the oldest unit; Kpf is the youngest. (After Mapel, W. J., G. S. Robinson, and P. K. Theobold, 1959. United States Geological Survey Map OM-191.)

2. Examine the outcrop pattern of the Cretaceous rock units shown in the lower left corner of the map. Note the V-shaped patterns formed where streams cut across these units. What is the structure of the Cretaceous units?

_____

3. The red lines on this map are faults. Note that many rock units are repeated as a result of the faulting. Based on their map pattern and their relationship to the folds located west of Little Rock in the Ouachita Mountains, what types of faults are exposed in the Ouachita Mountains? *Note:* the Ouachita Mountains

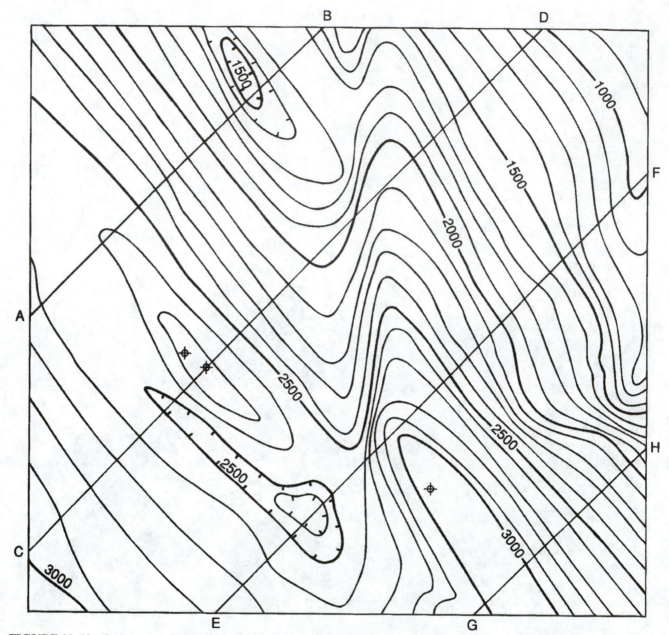

FIGURE 10–10   Structure contours drawn on the top of the Fall River Sandstone. For these cross sections, use graph paper with 10 divisions per inch. At this scale (1:96,000), 0.10 inch equals 800 feet. For 2× vertical exaggeration, 0.10 inch equals 400 feet. (After Mapel, W. J., G. S. Robinson, and P. K. Theobold, 1959. United States Geological Survey Map OM-191.)

lie south of the Arkansas River and north of the post-Paleozoic rocks of the Coastal Plain.

4. Describe the changes in structure you would cross if you made a traverse due north from Arkadelphia to the northern border of the map. Can you determine if the folds are plunging? If yes, in what direction do they plunge? Are the folds symmetric?

_____
_____
_____
_____
_____
_____
_____

5. How closely can you determine the age of the folding and faulting in the Ouachita Mountains?

_____

EXERCISE 10–4   Folds—Pittsfield Quadrangle, Massachusetts

Refer to the map in the appendix, page Ap2–21.

Note: Most of the rocks exposed in this map area are marbles. They were deeply
buried and became highly ductile during deformation. The black line that passes
through the D in Pittsfield is the axial trace of a fold.

1. What type of fold is the fold for which the axial trace is shown?_____

2. The red line that trends north–south across the map shows the axial trace of
   another fold. What type of fold is this? _____

3. Using a piece of tracing paper, make an overlay for this map and draw axial
   traces for the other folds shown on the map.

4. How can you explain the two sets of fold axes?

_____
_____

5. Which fold set is older? _____

EXERCISE 10–5   Folds—Williamsport Quadrangle, Pennsylvania

Refer to the geologic map in the appendix (page Ap2–33). Scale 1:24,000

The topographic contour interval is 20 feet.

A cross section has been drawn across the map along the line X–Y, Figure 10–11.
Some of the rock units shown on the map have been lumped together in order to
simplify the cross section. In this section Sr and Sm, all units between Sb and Dmt,
and Dh and Dtr are combined.

1. Draw a cross section across the map along the line from the point marked 349
   to point 87 in Figure 10–11b. In what direction do the two large folds on this
   map plunge? Draw the following contacts: Dm/Don, Sb/Sm, Sr/St, and St/Oj.

_____
_____
_____

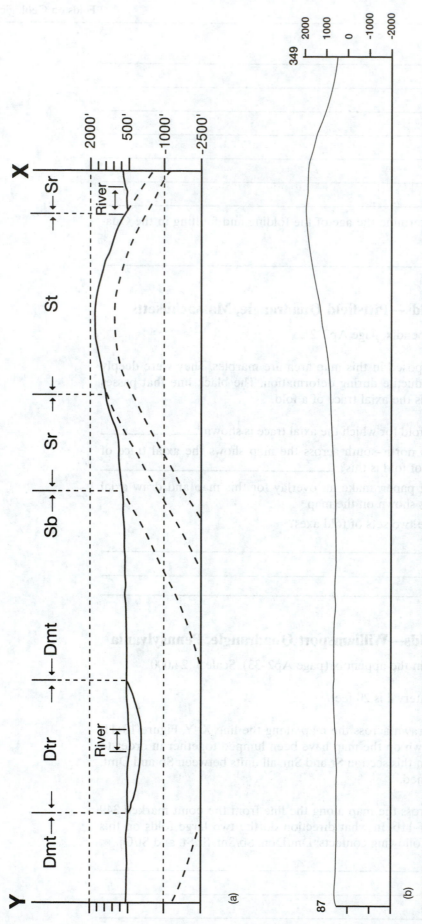

FIGURE 10–11 Cross section across the Bald Eagle Mountain on the Williamsport Region, Pennsylvania. (Based on Lloyd, O. B. Jr., and L. D. Carswell, 1981, "Groundwater resources of the Williamsport region, Lycoming County, Pennsylvania," Water Resources Report 51, Commonwealth of Pennsylvania Department of Environmental Resources.)

2. Explain the map pattern at point A. Why does Ojl occur as an isolated, elongate outcrop?

_____

_____

_____

_____

3. Why does the outcrop of Dh on the north and south limbs of this fold almost join at point B?

_____

_____

_____

_____

4. Explain the map pattern of Sm (the long, narrow connected strips) along an east–west line through point C.

_____

_____

_____

_____

5. Explain why the width of the outcrop belt of St varies so much.

_____

_____

_____

_____

EXERCISE 10–6   Folds—Salem Quadrangle, Kentucky

Refer to the geologic map in the appendix (page Ap2–27).

The structure contours on this map are drawn on the base of the Bethel Sandstone (Mcb). Negative values indicate elevations below sea level.

1. Is the area where IPca is exposed best described as a dome, basin, plunging anticline, or plunging syncline? _____

2. Are the rock units that are exposed outside the area bounded by faults folded?

_____

3. What are the approximate strike and dip of the rock units located outside the fault-bounded area?

_____

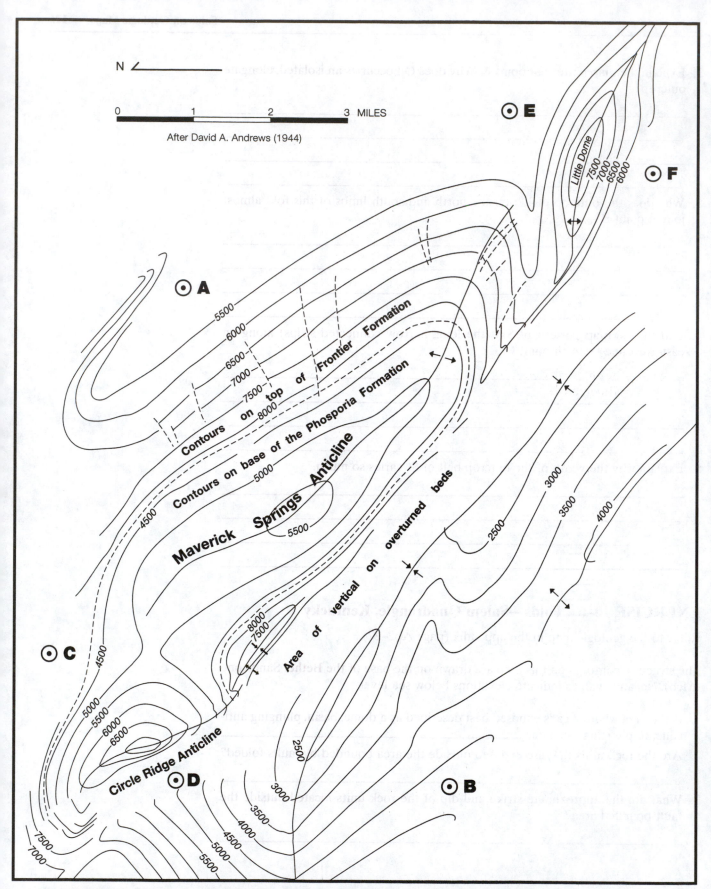

FIGURE 10–12 Structure contours drawn on top of the Frontier Formation except in the Maverick Springs and Circle Ridge anticlines, where the contours are drawn on the base of Phosphoria Formation. The structure contour interval is 500 feet. (After Andrews, D. A., 1944. Geologic and Structure Contour Map of the Maverick Springs Area, Fremont County, Wyoming, Oil and Gas Investigations Map 13, United States Geological Survey.)

EXERCISE 10–7   Folds—Maverick Springs Anticline, Wyoming

Refer to Figure 10–12.

1. Draw a cross section across the Maverick Springs anticline along the line A–B showing the shape of the Frontier and Phosphoria Formations. The structure contours show elevations above sea level on the base of the Permain age Phosphoria Formation and on the top of the Upper Cretaceous Frontier Formation. The stratigraphic interval between these two formations includes the entire Mesozoic section.

   A dashed line separates the portions of the map drawn on these two surfaces. The thickness of the stratigraphic section between the base of the Phosphoria and the top of the Frontier Formations is approximately 3,800 feet. Suggestion: draw the cross section using contours on the top of the Frontier Formation at the two ends of the cross section. Then use the thickness of the Phosphoria to estimate where the top of the Phosphoria will be below the top of the Frontier. Use dotted or dashed lines to show where you infer the position of the contacts to be across portions of the cross section where no control data are shown.

2. Draw cross sections across Little Dome and Circle Ridge anticlines along the lines C–D and E–F.

3. Compare the shape of Little Dome with that of the Maverick Springs anticline and Circle Ridge anticline.

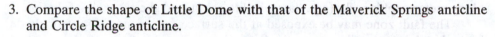

CHAPTER 11

Faults on Geologic Maps

Faults are breaks or zones across which portions of the crust have been displaced. Faults that form high in the crust tend to be zones of brittle deformation characterized by broken rock, called **breccia,** or zones of rock that have been ground down to a powder-like material, called **gouge.** When displacements occur deep in the crust, deformation takes place under temperature and pressure conditions that make the rock ductile. Rocks in these deep zones of deformation commonly exhibit a fabric in which the minerals are strongly aligned. Geologists call this mineral alignment **foliation.** Many metamorphic rocks (e.g., slate, phyllite, schist, and gneiss) exhibit foliation.

The fault zone may be exposed at the surface, but because the strongly deformed rock is susceptible to weathering and erosion, many fault zones are not exposed. Instead, soil or alluvium covers the bedrock. In such places, the presence of the fault can be detected by abnormal stratigraphic relationships. For example, part of a stratigraphic sequence may be repeated or absent. On geologic maps, faults are generally represented by lines that are heavier than those used for rock unit contacts. The fault type is indicated by use of special symbols (see the inside back cover).

The surface of the ground across recent faults may be displaced if the fault involves a vertical component of motion. In such instances, an escarpment, commonly called a fault scarp, marks the fault. In the basin and range province of western North America, deep basins mark the downthrown side of major faults.

The map pattern of contacts between rock units is usually offset by faults that cut across these contacts. If the strata cut by a steeply dipping fault are horizontal, the elevation of the contact will obviously be changed if there is a vertical component of movement on the fault. Locally, at least, this difference in elevation is a measure of the throw on the fault. If the strata cut by a steeply dipping fault are not horizontal, a more complex pattern may result. Unless the fault is recent, erosion will likely have lowered the level of the ground on the upthrown side of any fault. If the fault is an old one, the upthrown side of the fault may even be eroded to the same level as the downthrown side of the fault (Figure 11–1).

Over long periods, erosion generally removes escarpments formed at the time of displacement. For faults that formed many millions of years ago, there may be little or no relationship between the displacement and modern topography. Usually the crust on the upthrown side of a fault has been eroded more rapidly than that on the downthrown side. **Where the upthrown side of a fault is eroded more rapidly than the downthrown side, the position of contacts between dipping beds migrates in the direction of the dip of the contact.** This important concept is most useful in interpretation of maps showing faults (Figure 11–1).

FAULT NOMENCLATURE

1. Displacement.   A measure of the amount and direction of the actual movement that takes place when two fault blocks are displaced. *Syn.* slip.

114

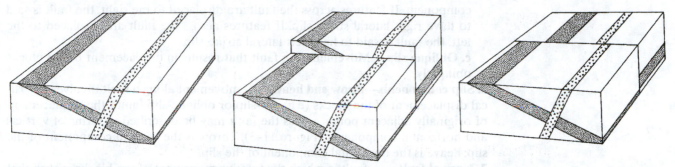

FIGURE 11–1 This sequence of block diagrams illustrates the changes that occur in the position of contacts as erosion lowers the upthrown block. In all cases the contact shifts—migrates—in the direction of the dip of the beds on the eroded upthrown block.

2. Fault trace. The line formed by the intersection of a fault with the ground surface. It is this trace that is shown on geologic maps.

3. Fault zone. A fault zone is the surface within which movement on a fault occurs. Commonly, this movement involves some thickness of rock (hence zone). Some fault zones are plane, but more often rock is disrupted, deformed, and distorted through a zone, meters and occasionally kilometers thick. Rock within the fault zone may be brecciated (broken into angular fragments) or crushed to form a powder called gouge. If the movement occurs under conditions favorable for recrystallization of the material in the fault zone, a very fine-grained, solid rock called mylonite may be formed.

4. Fault blocks. A fault may be thought of as a surface that locally divides the crust into two blocks. If the fault is vertical, the separate blocks are not given special names; but if the fault dips, the two blocks are distinguished from one another as the hanging wall block and the footwall block.

  a. Hanging wall block. The block of the crust that lies above a dipping fault.

  b. Footwall block. The block of the crust that lies beneath a dipping fault.

5. Slip. A measure of the amount and direction of the actual movement that takes place when two fault blocks are displaced.

  a. Dip slip. Movement in a fault directly up or down the dip of the fault (at right angles to the strike of the fault); see Figure 11–2.

  b. Strike slip. Movement in a fault that results in displacement of the blocks parallel to the strike of the fault. This causes displacement without a vertical

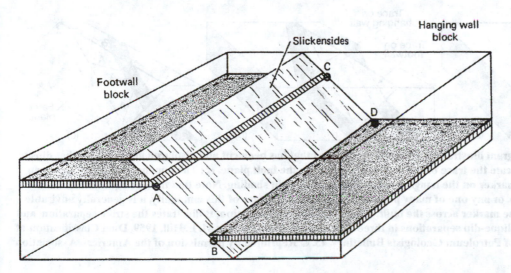

FIGURE 11–2 Block diagram shows a normal fault on which the hanging wall is down relative to the footwall. Slickensides indicate dip-slip movement. The distance A–B is called the displacement or slip. In this instance the displacement is equal to the dip separation.

component. If features across the fault are displaced to the right, the fault is said to have right lateral strike slip. If features across the fault are displaced to the left, the fault is said to have left lateral strike slip.

c. Oblique slip.   Movement in a fault that results in displacement of the blocks obliquely.

6. Slip components—throw and heave.   If movement along a fault involves a vertical displacement of the blocks (e.g., dip-slip or oblique-slip fault), the displacement of originally adjacent points across the fault may be described in terms of vertical and horizontal components (Figure 11–3). Throw is the vertical component of the slip; heave is the horizontal component of the slip.

7. Normal fault.   A fault in which the relative movement of the blocks is such that the hanging wall has moved down relative to the footwall.

8. Reverse fault.   A fault in which the relative movement of the blocks is such that the hanging wall has moved up relative to the footwall.

9. Stratigraphic throw.   Stratigraphic throw is the thickness of a section that is missing at any point along a fault. If the normal thicknesses of rock units in the map area are known, stratigraphic throw can be determined from a geologic map.

10. Separation.   The distance between markers (such as contacts between beds or dikes that are cut and displaced across a fault) measured in a specified direction relative to the fault plane. Because the amount and direction of slip on a fault are often unknown, it is important to recognize that apparent displacement of contacts observed in the field and on geologic maps and sections usually is not a measurement of slip or slip components (Figure 11–4). Designation of the apparent displacement

FIGURE 11–3   Throw and heave are illustrated for a normal fault with dip-slip displacement in this cross section drawn perpendicular to the fault.

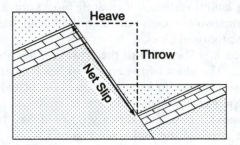

FIGURE 11–4   (a) Block diagram illustrating a dipping marker bed that has been cut and displaced along a fault that dips to the right. (b) Dashed lines indicate the trace of the displaced marker on the fault plane. (c) The hanging wall block has been removed, but the trace of the marker on the hanging wall is indicated by light shading. Note that this configuration could be the result of dip-slip, strike-slip, or any one of many possible oblique slips. Because of this ambiguity, it is generally advisable to describe the separation of the marker across the fault. C indicates the dip separation, B illustrates the strike separation, and the lines marked A indicate oblique-dip separations in three different directions. (After M. L. Hill, 1959, Dual Classification of Faults: American Association of Petroleum Geologists Bulletin, v. 43. © Reprinted by permission of the American Association of Petroleum Geologists.)

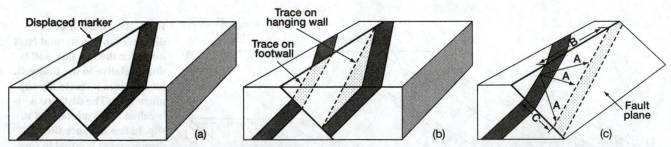

of markers across a fault as separations is a way of indicating that the person recording the measurements recognizes that these are not necessarily measurements of slip.

a. Strike separation.   The distance a stratum or other marker is displaced along the strike of a fault.

b. Dip separation.   The distance a stratum or other marker is displaced directly down the dip of a fault.

c. Oblique separation.   The distance a stratum or marker is displaced and measured in some specified direction.

d. Vertical separation.   The distance between two parts of a displaced stratum or other marker and measured in a vertical line (as in a well).

e. Offset.   The distance between two parts of a marker that has been displaced along a fault, measured perpendicular to the marker.

EXERCISE 11–1   Displacement and Separation

Schematic block diagrams and geologic maps (as they would appear if the upthrown side of each fault was eroded down to form a level surface) are shown in Figure 11–5. The blocks are shown displaced. From these it is possible to identify the actual displacement on the fault. For each case, identify the type of displacement (strike slip, dip slip, oblique slip) indicated by the block diagram, and the amount and direction (e.g., right lateral) of strike separation indicated.

a. Displacement: _____   Strike separation: _____
b. Displacement: _____   Strike separation: _____
c. Displacement: _____   Strike separation: _____
d. Displacement: _____   Strike separation: _____
e. Displacement: _____   Strike separation: _____
f. Displacement: _____   Strike separation: _____

CROSS-SECTION CONSTRUCTION IN FAULTED AREAS

As with other cross sections, start by preparing a topographic profile along the line you have selected for the cross section. Mark the position of contacts, including any faults that lie along the profile. If strike and dip data are available for contacts of rock units that cross the line, draw short lines below the profile indicating the direction and amount of dip. Examine the map pattern and symbols used to represent the fault. Faults that dip at high angles form straight lines where they cut across the topography; faults that have low dips form irregular patterns that weave in and out across the topography. Most geologic maps do not show dips for faults. You will have to interpret the pattern using the information about map patterns in the following sections.

HIGH-ANGLE FAULTS

Faults that are planar or near planar and have dips in the range of 50° to vertical are included in this discussion. Several distinctly different types of faults have dips in this range—including strike-slip faults, normal faults, and reverse faults. Of these, strike-slip faults are commonly vertical or so close to vertical that the dip cannot be distinguished from vertical on a geologic map. Both normal and reverse faults commonly have steep dips. In addition, portions of some thrust faults have steep dips.

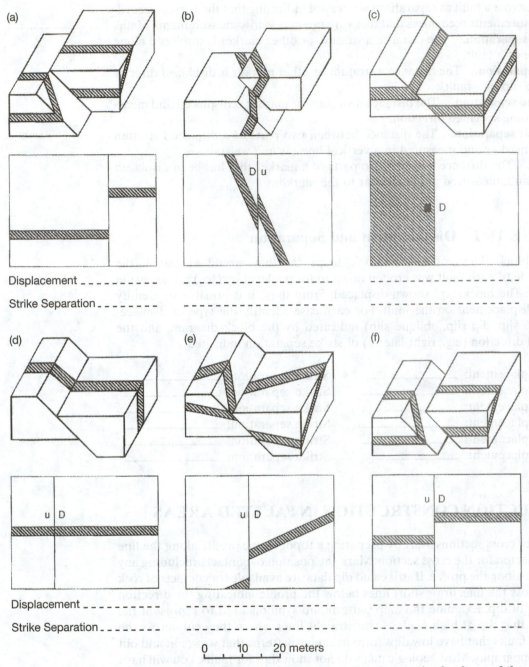

FIGURE 11–5 Block diagrams and maps showing displaced marker after erosion has lowered the upthrown side of the fault to the level of the downthrown block.

Patterns of High-Angle Faults on Geologic Maps

The most distinctive feature of the appearance of steeply dipping faults on geologic maps is their "straight line" trace. Any vertical plane (e.g., fault, dike, or bed) will form a straight trace across a map despite topographic irregularity. If a fault is not vertical but is approximately plane, its strike and dip can be determined using three-point solutions, described earlier.

Certain map patterns (Figure 11–6) commonly occur where high-angle faults are present.

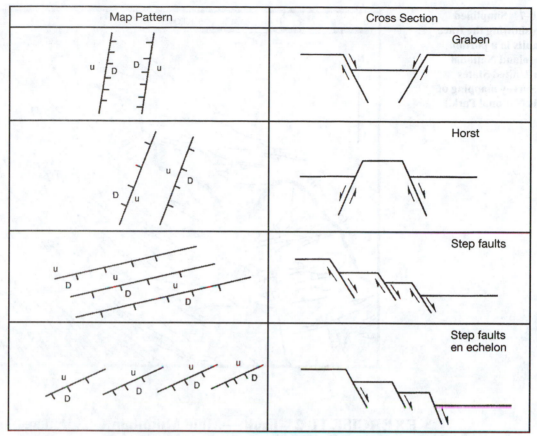

FIGURE 11–6   Schematic map patterns and corresponding cross sections for some common fault patterns associated with high-angle faults.

Grabens.   A strip of crust bounded by normal faults that dip toward one another. The central strip of rock is down relative to that outside the graben. Although all grabens develop as a result of crustal extension, they may form in a variety of structural settings. The grabens along the southeastern edge of the Colorado River in Canyonland National Park (Figure 11–7) formed as a result of slip over a layer of salt cut by the Colorado River as it entrenched its channel. Salt is a weak rock that has low resistance to shear. Once the river had cut its channel to the salt, lateral support for the layers over the salt decreased, and they slipped toward the river. Major systems of grabens, half-grabens, and step faults lie along continental margins where continents have rifted. A number of these features are present in the Appalachian Mountains (Figure 11–8). Many more grabens are present in the subsurface beneath Mesozoic and Cenozoic sediments along the Atlantic continental margin. On a much smaller scale, high-angle normal faults forming grabens and horsts are also common features in the rocks forced into domal features by the rise of plugs of salt from below (Figure 11–9). Similar features form over some igneous intrusions.

Horsts.   A strip of crust bounded by normal faults that dip away from one another. The strip is upthrown relative to adjacent crust.

Step Faults.   Where several normal faults are parallel or subparallel to one another, dip in the same direction and have similar displacement, a steplike pattern may be created (Figure 11–10). Small faults parallel to a larger fault are called synthetic faults.

En Echelon Faults.   This term refers to a pattern characterized by systematic offset of short, parallel faults.

FIGURE 11–7 Simplified map pattern showing the trace of normal faults in a portion of the Canyonland National Park. (After United States Geological Survey mapping of Canyonlands National Park.)

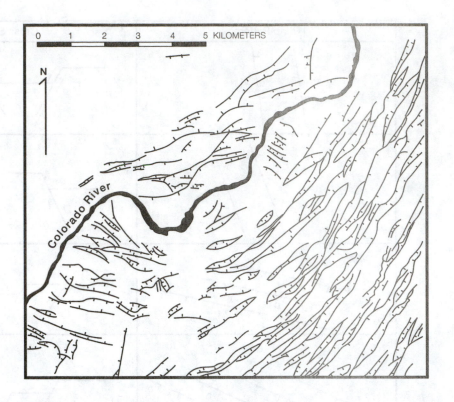

EXERCISE 11–2   Faults—Mule Mountain Quadrangle, Arizona

Refer to the map in the appendix, page Ap2–21.

1. Describe the strike and dip of the Cretaceous formations (use the symbols on the map).

   _____

2. A fault trends approximately east–west near the bottom of the map. Without referring to the symbols, how can you determine which side of this fault is upthrown?

   _____

3. Several faults trend north–northeast across the map. What evidence from the map indicates that these faults have steep dips? _____

   _____

4. Without referring to the symbols, how can you determine which side is upthrown?

   _____

5. What would you need to know in order to calculate the throw on the fault labeled A?

   _____

   _____

EXERCISE 11–3   Faults—Brownwood and Llano, Texas

Refer to the map in the appendix, page Ap2–7.

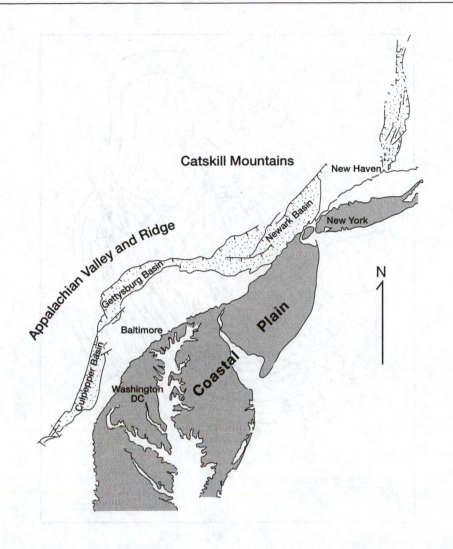

FIGURE 11–8 A number of large grabens and half-grabens lie within the Appalachian Mountains along the east coast of North America. Most of these large basins formed during the early stages of the opening of the Atlantic Ocean in the Mesozoic. The basins contain sedimentary rocks eroded from the Appalachians. (After the Tectonic Map of the United States, © AAPG, reprinted by permission of the American Association of Petroleum Geologists.)

Note: The tick marks indicate the direction of dip of the faults.

1. What types of faults are shown on this map? _____

2. What is the map evidence to support your answer to question 1?

_____

_____

3. Assume that all of the faults are of the same age.
   a. What is the youngest formation cut by faults? _____
   b. What is the oldest formation not cut by faults? _____

4. Where on the map do you see:
   a. Step faults _____

   _____

   b. Grabens _____

   _____

   c. Horsts _____

   _____

FIGURE 11–9   Fault pattern over the Hawkins salt dome in Texas. (From Wendlandt, E. A., 1951. "Hawkins Field, Wood County, Texas." Austin, University of Texas Publication no. 5116.)

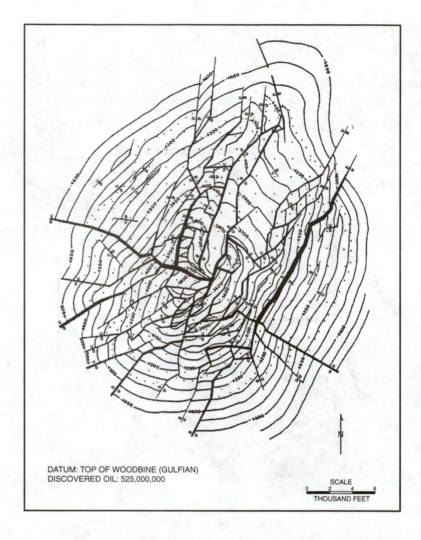

DATUM: TOP OF WOODBINE (GULFIAN)
DISCOVERED OIL: 525,000,000

SCALE

THOUSAND FEET

FIGURE 11–10   Step fault pattern of normal faults. (After Shoemaker, E. M., 1956, Geologic Map of the Roc Creek Quadrangle, Colorado, United States Geological Survey Map GQ 83.)

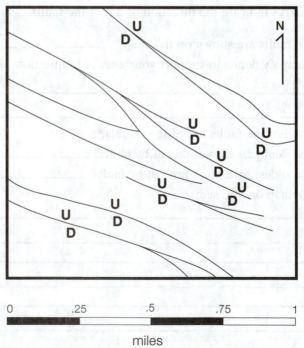

miles

EXERCISE 11-4   Faults—Grand Canyon, Arizona

Refer to the map in the appendix, page Ap2–13.

1. What types of faults are present in this region? _____

2. What is the relationship between the Bright Angel and Phantom Ranch faults and topographic features? _____

_____

3. Have streams formed along all faults? _____

4. Cite examples of faults that cut across ridges. _____

_____

5. Which side of the Bright Angel fault is upthrown? _____

6. How could you determine which side is upthrown without referring to map symbols? _____

_____

7. Examine the Cremation fault in the southeastern part of the map. Which side of the fault is upthrown? _____

8. What is the stratigraphic throw (express in terms of the units that are missing across the fault)? _____

9. Is there evidence that some of the faults are older than others? If yes, what is that evidence?

_____

_____

_____

EXERCISE 11-5   Faults—Atkinson Creek Quadrangle, Colorado

Refer to the geologic map in the appendix, page Ap2–5.

1. What is the direction of dip of the faults? _____

   What evidence supports this conclusion? _____

2. How could you determine the downthrown side if the faults were not labeled by using stratigraphy? _____

3. Using structure contours? _____

_____

_____

4. What is the direction of dip of the formations in the northern part of the map (north of the faults)? (note that Jmb lies above Jms) _____

   What evidence did you use? _____

5. What is the throw measured on the Entrada Formation across the two longest fault zones? (*Hint:* Use the structure contours to answer this question.) The heavy dashed structure contour in the southwestern part of the map is 6,000 feet.

   a. Northern fault: _____

   b. Southern fault: _____

EXERCISE 11–6   Faults, Pine Mountain Quadrangle, Colorado

Refer to the geologic map in the appendix, page Ap2–21.

1. Which direction does the fault that trends NW to SE across the map dip? (*Hint:* Use the law of Vs, discussed in the section on geometry of sedimentary rock bodies. Use the V-shaped contact in section 19.) _____
   What evidence supports your answer? _____
   _____
   _____

2. What is the angle of dip of the fault? _____

3. In the area northeast of the long fault (northeast part of section 19), what is the direction of dip of the Ʀc/Jw contact? _____

4. In the area southwest of the long fault (in section 30), what is the direction of dip of the Ʀc/Jw contact? _____

5. Based on the shape of the Vs formed along this contact, how does the dip of the Ʀc/Jw contact in section 30 compare with that of the fault? _____
   _____

6. Measure the angle of dip using the structure contours (note that the first dashed contour southwest of the fault is the 8,000-foot contour). _____

7. Draw a schematic cross section across the fault southwest of the southeast corner of section 19.

EXERCISE 11–7   Faults—Salem Quadrangle, Kentucky

Refer to the geologic map in the appendix, page Ap2–27.

1. What evidence indicates that these faults are nearly vertical? _____
   _____

2. The structure contours are drawn on the base of the Bethel Sandstone (Mcb). What type of fault-bounded structure runs NE–SW across the map? _____
   _____

3. Describe the structure indicated by the structure contours. _____
   _____

4. Which sides of the faults labeled A B C are upthrown?
   A: _____
   B: _____
   C: _____

5. Which faults have the greatest throw? _____
   Which fault has the least throw? _____

6. Some of the structure contours terminate against the faults. Others are closed to form complete ovals. What is the shape of the marker on which the structure contours are drawn where the marker is cut by a fault?
   _____
   _____
   _____
   _____

7. Locate this map on the tectonic map of the United States. The faults shown on this map are part of a much larger system of faults. What relationship do the trends of these faults bear to that of the Mississippi Embayment?

_____

_____

_____

8. What age relationship do these faults bear to the rocks in the Mississippi Embayment?

_____

_____

EXERCISE 11–8   Faults—Kilauea Crater Quadrangle, Hawaii

Refer to the map in the appendix, page Ap2–19.

1. What types of faults are shown on this map? _____

2. Where, if anywhere, do you see evidence of step faults or a graben?

_____

_____

_____

3. In general, which side of the faults is downthrown?

_____

PATTERNS ALONG STRIKE-SLIP FAULTS

Some strike-slip faults are localized breaks in crustal rocks that form where the lateral movements of sedimentary layers are not uniform. Many such faults are tears that develop along folds or along thrust faults. Such faults are generally of limited length, and terminate at depth in a layer of weak rock such as shale or salt, but tear faults can also form in thin sheets of crystalline rocks. Other strike-slip faults appear to penetrate the thickness of the crust and possibly the entire thickness of the lithosphere, forming plate boundaries. The San Andreas fault of California is among this latter group. These faults form broad zones commonly containing numerous faults that are parallel or subparallel to one another. Many of the faults in these zones are nearly vertical and exhibit strike-slip displacement, but thrust faults and folds also occur within the zones (Figure 11–11). The thrusts result from lateral compression across the fault zone or from within the zone. Individual faults within many major strike-slip fault zones are discontinuous, and offsets in displacement occur where one fault terminates and another begins. Depending on the geometry of the offsets where individual faults terminate or overlap, a depression or a bulge may form at the ground surface (Figure 11–12). If the depression contains a standing body of water, it is called a sag pond.

LOW-ANGLE FAULTS

Thrust faults are defined as low-angle reverse faults, but the dip of the fault plane may vary from nearly vertical to horizontal. Small thrust faults may occur on either the forelimb or the backlimb of folds (Figure 11–13a and b). The fault zone may con-

(a)

(b)

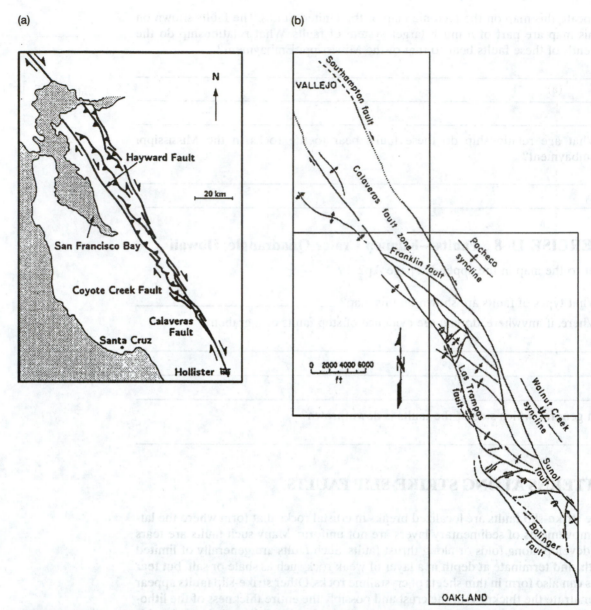

FIGURE 11–11   (a) Thrust faults are associated with the Hayward and Calaveras strike-slip faults. (After mapping by Aydin and Page). (b) Part of the Calaveras fault zone of California showing bifurcations along the fault and folds caused by movements along the fault zone. (After R. B. Saul, 1967, "The Calaveras Fault Zone," Mineral Information Service, vol. 20, no. 3.)

sist of a single sharply defined plane or rupture; but near its leading edge the fault commonly splits, forming a series of slices or imbricate faults (Figure 11–13c).

On most thrusts, the lateral component of displacement is large relative to the vertical component. Along the great thrust faults, some of which extend for hundreds of kilometers, huge slabs of the upper part of the crust have moved great distances. Along some thrust faults in orogenic belts, displacements amount to tens, even hundreds, of kilometers. Such faults are rarely planar. In some places, the plane of the thrust fault may lie within a bedding plane. Usually these sections of the fault plane lie with weak stratigraphic units such as shales or evaporites. In other places, the

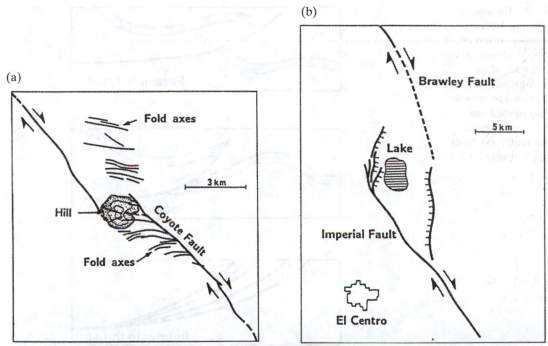

FIGURE 11–12 Uplifted and downwarped areas occur along some strike-slip fault zones. (a) A hill lies between two segments of Coyote Creek fault in southern California. (After Sharp, R. V., and M. M. Clark, 1972. Geologic evidence of previous faulting near the 1968 rupture on the Coyote Creek fault, in The Borrego Mountain earthquake of April 9, 1968: United States Geological Survey Professional Paper, no 787.) (b) A lake fills a depression between Brawley and Imperial faults near El Centro, California. (After Johnson, C. E., and D. M. Hadley, 1976. "Tectonic Implications of the Brawley Earthquake Swarm, Imperial Valley, California, January 1975," *Seismological Society of America Bulletin,* vol. 66, no. 4, pp. 1,132–1,144. © 1976 Seismological Society of America.)

thrust may cut up through the stratigraphic section forming a **ramp** (Figures 11–13e and d) where it moves from one weak zone to another. Initial dips at ramps are generally on the order of thirty degrees. If the thrust develops early in the history of deformation of a region, the plane of the thrust may be folded by later deformation.

One result of the geometry of thrust faults is that their trace may be parallel to rock unit contacts in some places and cut across them in others (Figures 11–14 and 11–15). Where the fault plane is vertical, the fault trace may be an essentially straight line on the map, and the map pattern may resemble that of a high-angle fault. Where thrusts are flat or nearly flat, the trace of the fault may have a highly irregular pattern where it cuts through the topography. If the fault plane is nearly horizontal and lies close to the ground surface, erosion may cut through the thrust plane, exposing rocks that lie beneath the fault. Geologists refer to these features as **windows.** Erosion of the fault plane may also leave pieces of the thrust sheet isolated from the main part of the thrust sheet. These remnants are **klippes** (Figure 11–16).

Not all faults with low dips are thrusts. Normal faults may curve and flatten at depth. Such normal faults have steep dips near the surface, but the fault plane curves beneath the surface and may eventually become horizontal, flattening out in a bedding plane. In areas of complex geology, older faults may be folded or cut by faults formed during a younger deformation. Faults in the Orem Quadrangle (Figure 11–17) are examples of this. In this area, the Baldy Thrust is cut by younger high-angle faults.

FIGURE 11–13   Geometry of thrust faults. (a) A thrust cuts through the forelimb of an anticline. (b) A thrust cuts across the backlimb of an anticline. (c) Imbricate thrust faults. (d) A pop-up formed by thrusts that dip toward one another. (e) A ramp formed where a thrust fault rises from one stratigraphic level to another.

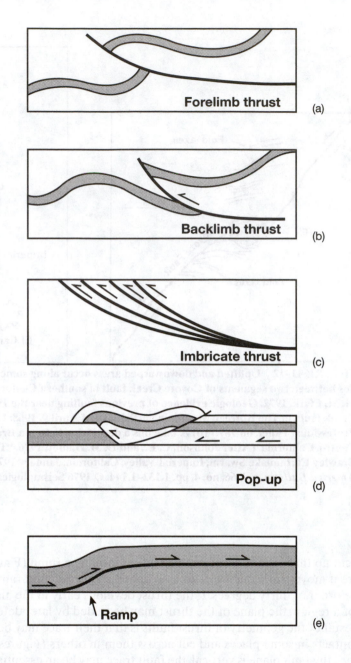

EXERCISE 11–9   Faults—Smoky Mountains, Tennessee and North Carolina

Refer to the map in the appendix, page Ap2–29.

1. What type of fault is the Great Smoky fault? _____

2. In which direction does the Great Smoky fault dip (along Chilhowee Mountain)?

    _____

    What evidence do you see for this at the point labeled (A)?

    _____

    _____

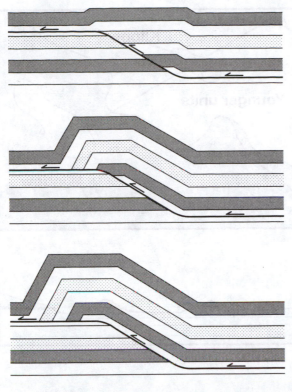

FIGURE 11–14   Stages in the growth of a fault-bend fold (After Suppe, J., and J. Namson, 1978. "Fault-bend Origin of Frontal Folds of the Western Taiwan Fold-and-Thrust Belt," *Petroleum Geology of Taiwan,* **no. 16, pp. 1–18.)**

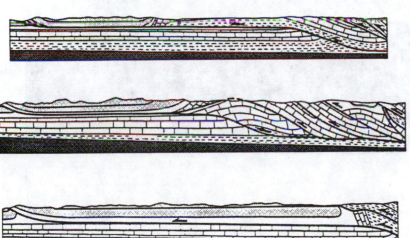

FIGURE 11–15   Cross sections along lines across the Pine Mountain overthrust plate. (After Harris, L. D., and R. C. Milici, 1977, "Characteristics of thin-skinned style of deformation in the southern Appalachians, and potential hydrocarbon traps." United States Geological Survey Professional Paper 1018.)

3. A fault lies along the Little Pigeon River [northeast corner of map; point (B)]. What type of fault is this fault?

_____

4. Is the separation along the fault at Little Pigeon River right or left lateral?

_____

5. What is the strike separation of the Great Smoky fault measured along the Little Pigeon River?

_____

6. What is the structural feature labeled (C) that is located in Tuckaleechee Cove?

_____

FIGURE 11–16 Map and cross section showing a klippe and a window formed along a thrust fault.

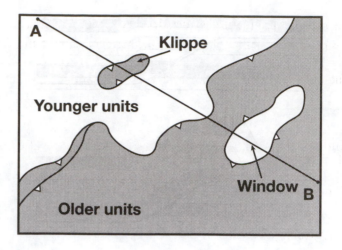

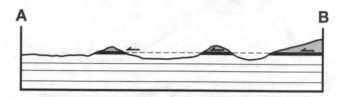

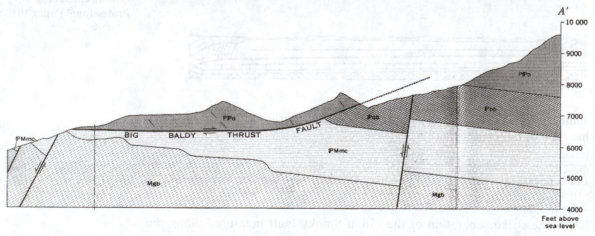

FIGURE 11–17 Portion of the Orem Quadrangle geologic map and a cross section across the Big Baldy thrust fault. (After Baker, A. A., 1964. United States Geological Survey Map GQ-241.)

7. What type of structural feature is Tuckaleechee Cove? _____

8. In what direction was the compression acting that caused the faults in this region? _____

9. Explain how the faults located north of point (D) formed.

   _____

   _____

10. What is the age of the faults north of point D relative to the Greenbrier Fault?

    _____

11. Point (E) is located along the axial trend of what type of fold?

    _____

12. Explain the structure of the fold located at point (F).

    _____

    _____

13. If the fault exposed around Tuckaleechee Cove is the Great Smoky fault, what is the minimum amount of horizontal displacement on the Great Smoky fault?

    _____

14. What is the stratigraphic throw (express in terms of the rock units that are missing) along the fault the surrounds Wear Cove? _____

15. How would you be able to identify the fault type for the faults around Tuckaleechee and Wear Coves even if they had no symbols on them?

    _____

    _____

EXERCISE 11–10   Faults—Duffield Quadrangle, Virginia

Refer to the map in the appendix, page Ap2–11.

1. If there were no symbols on the map, how would you be able to determine that the large fold that lies in the middle of this map is a syncline?

   _____

2. In which direction does the syncline plunge? _____

3. Is the fold symmetrical or asymmetrical? _____
   What is the evidence for your answer? _____

4. Where do you find evidence of a small fold on one limb of the large fold?

   _____

5. Three major faults, the Clinchport, Hunter Valley, and the Red Hill (located south of the Pattonsville Branch in the upper left part of the map) are on this map.

   a. All three of these faults are of what type? _____

   b. Which of these faults has the lowest dip? _____

   c. Which of these faults has the steepest dip? _____

6. In which direction do the faults dip? _____

7. What rock units are cut out along each of the faults?

   a. Clinchport _____

   b. Hunter Valley _____

   c. Red Hill _____

8. If we knew the thickness of each of the stratigraphic units, we could calculate the stratigraphic throw on each fault. Which fault has the largest stratigraphic throw? _____

9. Draw a freehand sketch of the structure you would expect to see along the line A–B.

10. Draw a cross section from the surface to sea level along the line C–D.

EXERCISE 11–11   Faults—Calgary, British Columbia and Alberta

Refer to the map in the appendix, page Ap2–9.

1. Describe the structure of the portion of this map that lies between Big Prairie and the McConnell thrust. Include in your description the number of faults you would cross, the types of faults present, the direction of dip of those faults, and the evidence you see that indicates the direction of dip.

_____

_____

_____

_____

2. What types of folds do you see in the belt described in question 1, known as the Foothills Belt?

_____

3. Are most of these folds of the same type? _____

4. Are the folds and faults related? If yes, in what way are they related?

_____

5. Do you see any klippes or windows in the foothills belt? _____

6. Compare the age of the rocks on the western side of each fault with those on the eastern side.

   a. Is the pattern consistent? _____

   b. On which side is older rock exposed? _____

   c. Explain your answer to b.

_____

_____

7. How do the map patterns of the fault west of the McConnell thrust differ from the patterns of the faults in the Foothills Belt?

_____

_____

8. What does this indicate about the dip of the faults?

_____

9. What are the ages of the rocks west of the McConnell thrust? _____

10. How does the age of the rocks in the Main Ranges west of the McConnell thrust differ from that of the rocks in the Foothills Belt east of the McConnell thrust?

_____

_____

11. Examine the structure of the **Panther River Culmination**. A number of thrusts are associated with the culmination. In terms of its gross structure, what type of fold is the culmination?

_____

12. In which direction is the syncline that lies along the western edge of the culmination overturned? _____

13. Explain why this fold is overturned in the direction you cited in question 12.

_____

_____

EXERCISE 11–12   Faults—Paradise Peak Quadrangle, Nevada

Refer to the map in the appendix, page Ap2–25.

1. Are gro and gd younger or older than the Paradise thrust? _____

2. What is the evidence for your answer?

_____

3. What rock unit makes up most of the Paradise thrust sheet? _____

4. Is Tvu younger or older than the Paradise thrust? _____

5. How closely can you determine the age of the Paradise thrust?

It is older than   _____

It is younger than   _____

6. What is the name of the structural feature surrounded by thrust faults in sections 17, 20, 7, and 8 in the southeastern section of the map?

_____

7. What is the name of the structural features surrounded by thrust faults in section 26 near the bottom of the map? _____

8. Is the South thrust younger or older than the Paradise thrust? _____

9. Why is the syncline in section 13 (near top center of map) overturned?

_____

_____

10. In which direction was the Paradise thrust sheet moving when it was emplaced? _____

11. What evidence do you find on the map to support your conclusion in question 10?

_____

_____

_____

_____

Igneous and Metamorphic Rocks

Igneous rocks include both plutons, which cool and solidify below the ground surface, and volcanic rocks, which reach the surface and take the form of lava flows or pyroclastic deposits consisting of fragments that drop from the air. Many of the pyroclastic deposits are sheetlike and may resemble sedimentary rocks both in the field and on geologic maps.

Metamorphic rocks are either transformed sedimentary or igneous rocks. Although most metamorphic rocks exhibit distinctive textures, their appearance on maps closely resembles the igneous or sedimentary bodies from which they originated. Metamorphism of sedimentary rocks altered by their proximity to igneous intrusions show up on geologic maps only if the map contains **isograds,** lines identifying the limits of certain key index minerals that indicate the degree of metamorphism, or some shading effect.

The igneous and metamorphic rocks exposed in many shield areas were metamorphosed at great depth in the crust. Some of these crystalline complexes formed at depths of 10km to 20km. At these depths, the temperature and pressure are so great that even plutonic rocks such as granite may become ductile. Partial melts are also common under these circumstances. If such rocks are subjected to later compression while they are ductile, complex deformational patterns may evolve as they have in the metamorphic rocks of Scotland (Figure 12–1) and in the Canadian Shield (Figure 12–2).

APPEARANCE OF PLUTONS ON GEOLOGIC MAPS

Generally, plutons have distinctive patterns on geologic maps. It is easy to spot discordant plutons because the boundaries of these plutons cut across other rock bodies or structural features that were present at the time of intrusion. Because their boundaries coincide with the boundaries between sedimentary layers or other contacts that were present before the intrusions, concordant plutons have less distinctive map patterns. However, in many cases, even the boundaries of concordant plutons locally cut across country rock or are the source of small discordant intrusions that extend out into the country rock. If the map pattern does not make identification of igneous rocks obvious, examine the explanation on the map. Igneous rocks are commonly separated from the sedimentary rocks and placed at the bottom of the stratigraphic column. On many maps, igneous bodies are differentiated from sedimentary rocks by the use of distinctive map patterns consisting of randomly placed (v or =) symbols.

NOMENCLATURE AND CLASSIFICATION OF INTRUSIONS

Large plutons, especially those that have not been strongly deformed, may appear as oval masses; but many also exhibit irregular shapes (Figure 12–2). For those with more clearly defined shapes, the forms of the pluton and its structural relationship to

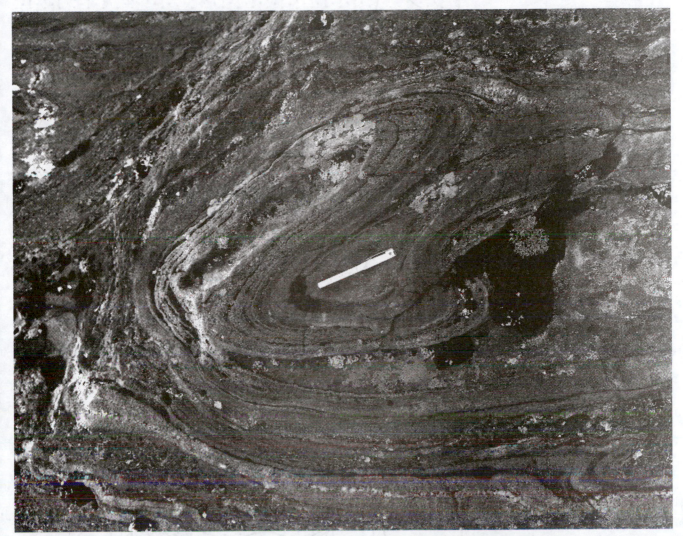

FIGURE 12–1 In orogenic belts, metamorphic rocks commonly exhibit complex fold patterns such as the one illustrated here from Scotland. The refolded folds seen here in outcrop may also occur on much larger scales that would show on geologic maps.

the enclosing rock are used for definition and classification (Figure 12–3). The more regular-shaped plutons may be defined and classified as outlined in the following sections (after a classification by R. A. Daly, 1933).

Concordant Plutons

Concordant plutons are those injected along planes of stratification, layering, or schistosity.

1. Sill. Sheetlike body with large lateral extent relative to its thickness. Transitions occur between skills and laccoliths. These are termed multiple if there are successive injections of one kind of magma, and composite if there is a compound intrusion of successive injections of more than one kind of magma.

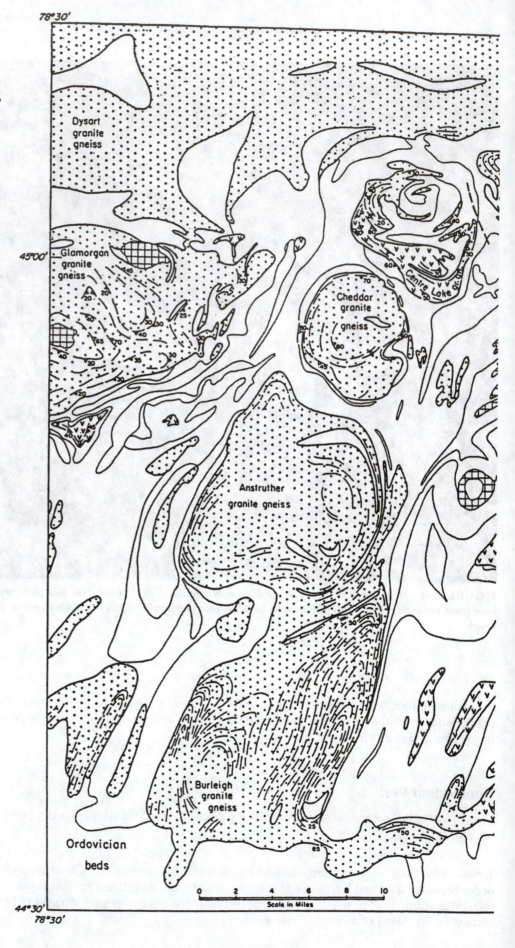

FIGURE 12–2 Geologic map of a portion of the Canadian Shield showing several large granitic intrusives. (From "Granite Emplacement with Special Reference to North America," *Geological Society of America Bulletin,* vol. 70, pp. 671–747, A. F. Buddington. Reproduced with permission of the publisher, the Geologic Society of America, Boulder, Colorado, USA. Copyright © 1959 Geological Society of America.)

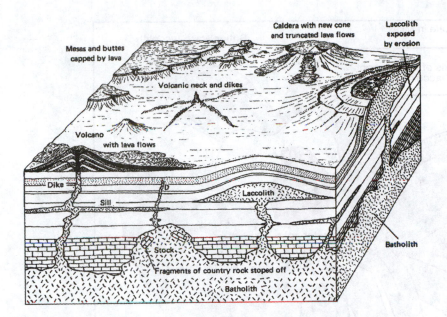

Caldera with new cone
and truncated lava flows

Laccolith
exposed
by erosion

Mesas and buttes
capped by lava

Volcanic neck and dikes

Volcano
with lava flows

Dike

Sill

D

Laccolith

Batholith

Stock

Fragments of country rock stoped off

Batholith

FIGURE 12–3   Block diagram illustrating some of the common intrusive and extrusive forms. (Courtesy of Frederick Young.)

2. Laccolith.   A planoconvex or doubly convex lens, flattened in the bedding plane of the invaded formation. The roof is arched over laccoliths (Figure 12–4b). These may also be multiple and composite intrusions (Figure 12–4a).

3. Lopolith.   A lenticular concordant intrusive mass in which the thickness is approximately one-tenth to one-twentieth of its width. The central portion is concave upward (Figure 12–4c).

4. Phacolith.   Intrusion of lensoid shape in cross sections, located at the hinge of folds. It may extend parallel to the axis of the fold.

Discordant Plutons

Discordant plutons are those injected across planes of stratification or schistosity.

1. Dike.   Injected body with parallel or subparallel walls that is narrow relative to its lateral extent.

2. Dike swarm.   Many dikes of similar trend or orientation occurring together.

3. Intrusive vein.   When the path of a discordant injected body is less regularly defined than is true of dikes, the wavy threadlike protrusion is called a vein.

4. Apophyses or tongues.   Dikes or veins that can be traced to larger intrusive bodies as the source of magmatic supply.

5. Ring dike.   A dike of arcuate to circular outcrop (Figure 12–4d).

6. Cone sheet.   A dike of arcuate outcrop and regular inclination toward a focus.

7. Volcanic neck.   Solid lava occupying a volcanic vent.

8. Batholith.   "A batholith is a stock-shaped or shield-shaped mass intruded as the result of the fusion of older formations. On the removal of its rock cover and continued denudation, the mass either holds its diameter or grows broader to unknown depth" (Suess, 1918). (Usage confines the name *batholith* to masses of more than 100 square kilometers' surface exposure.)

9. Stocks and bosses.   These two terms are used almost synonymously. They refer to masses similar to batholiths, except that they are smaller. Bosses are stocks of nearly circular ground plan.

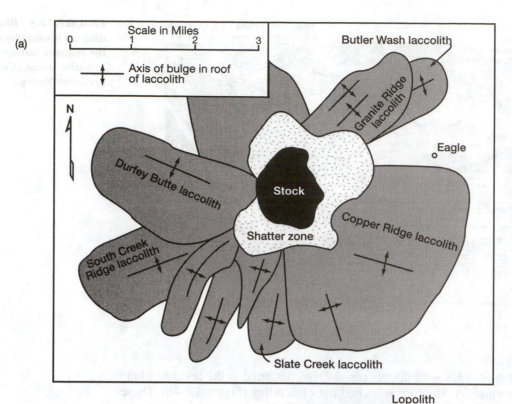

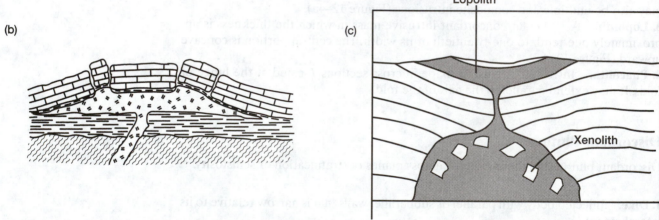

FIGURE 12–4 **(a) Map of the laccoliths and a central stock in the Henry Mountains, Utah. (After Hunt, C. B., Paul Averitt, and R. L. Miller, 1953. "Geology and Geography of the Henry Mountains Regions, Utah," United States Geological Survey Professional Paper 228. (b) Cross section of a laccolith. (After Hunt, 1953.) (c) Schematic cross section across a lopolith.**

EXERCISE 12–1   Plutons—Danford Quadrangle, Maine

Refer to the accompanying sketch map based on the Danford Quadrangle (Figure 12–5).

Scale: 1:24,000

Stratigraphic Units

db		Basaltic dike
Dg	Devonian	Granite and quartz monzonite
Ss	Silurian	Slate

(d)

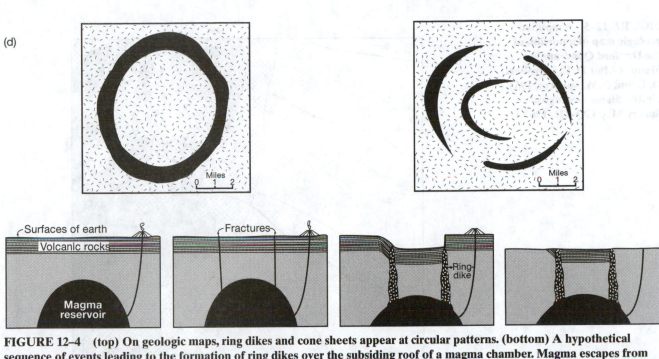

FIGURE 12–4   (top) On geologic maps, ring dikes and cone sheets appear at circular patterns. (bottom) A hypothetical sequence of events leading to the formation of ring dikes over the subsiding roof of a magma chamber. Magma escapes from the magma chamber (far left). Fractures form as subsidences start (left center). As subsidence continues, magma intrudes the fractures forming a ring dike (right center). Finally erosion bevels the region to a more or less flat surface (far right). (After Billings, M. P., 1972. *Structural Geology,* Prentice-Hall Inc.)

Sq	"	Quartzite, conglomerate, slate
Sss	"	Slate and siltstone
Os	Ordovician	Slate

1. If this map had no symbols or explanation, how might you interpret the map pattern relationship between Dg and the other units? Give more than one interpretation.

_____

_____

2. What does the dot pattern that is parallel to the border of Dg represent?_____

_____

3. Describe the structure of the metasedimentary rocks of this area. _____

_____

4. What is the relative age of Dg, db, and Sss? How did you determine these relative ages?

_____

_____

5. A zone of hills occurs in the area underlain by metasedimentary rocks near the contact zone of the pluton. How might you explain the presence of hills in this location?

_____

_____

_____

FIGURE 12–5 Simplified geologic map of a portion of the Danford Quadrangle, Maine. (After D. L. Larrabee, D. L. and E. W. Spencer, 1963. United States Geological Survey Map GQ-221, 1963.)

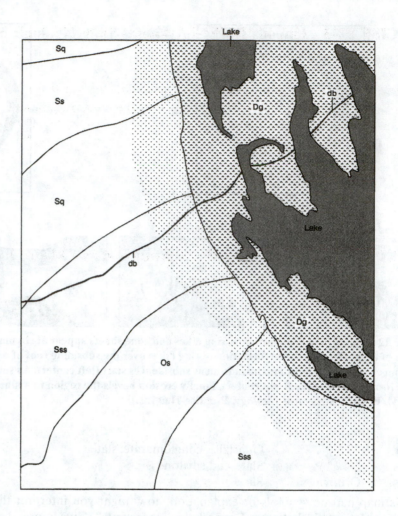

EXERCISE 12–2 Igneous Rocks—Paradise Peak Quadrangle, Nevada

Refer to the map in the appendix, page Ap2–25.

Plutonic and volcanic rocks shown on this map are Pg (Greenstone, a metabasalt), mz (monzonite), di (diorite), gd (granodiorite), gry (granite), gro (porphyritic granite), gr (red) (dikes, sills, and plugs of andesite and rhyolite), Tul (volcanic rocks), Tvu (volcanic rocks), and Qti (intrusive rocks).

1. How can you determine from the map that gr (red intrusives) are younger than gd? Cite evidence. _____

2. Are the red intrusive rocks younger or older than gro? _____
   Cite evidence. _____

3. Is gr younger or older than the Paradise thrust? _____
   Cite evidence. _____

4. Is gd younger or older than the Paradise thrust? _____
   Cite evidence. _____

5. Is gro younger or older than the Paradise thrust? _____
   Cite evidence. _____

EXERCISE 12–3   Igneous Rocks—Arkansas State Geologic Map

Refer to the map in the appendix, page Ap2–3.

1. Locate the igneous intrusions shown on this map. What age are they? _____

2. How can you explain the observation that rocks that are younger than the intrusions surround some of these intrusions located in the Coastal Plain?

_____

_____

_____

EXERCISE 12–4   Igneous Rocks—Brownwood and Llano, Texas

Refer to the map in the appendix, page Ap2–7.

1. What type of rock is €tm?_____

2. What name applies to the structure of the p€tm?

_____

_____

EXERCISE 12–5   Igneous Rocks—Jemez Mountains, New Mexico

Refer to the geologic map of the Jemez Mountains in the appendix (page Ap2–17). During the Tertiary and Quaternary the Valle Grande volcanic center has been the site of some of the largest and most violent eruptions in North America. The center is located on the western margin of the Rio Grande Rift zone, a large graben that extends north–south across New Mexico and into Colorado.

1. Locate the oldest rocks shown on this map. Where are they located in the topography (e.g., ridge tops, valley floors, etc.)?

_____

_____

2. How can you explain the topographic location of the oldest rocks?

_____

_____

3. The large circular feature near the center of this map is a caldera. Compare the shape and topographic features in this caldera with those in the caldera at Kilauea Crater.

_____

_____

4. Describe the shape of the features formed by Qvvf.

_____

_____

5. What is the relative age of the features described in question 4 and formation of the caldera?

_____

_____

6. Compare the fault patterns formed near the center of the caldera with those shown in Figure 11–9 (developed over a salt dome). What caused the faults in the caldera to form?

_____

_____

_____

_____

EXERCISE 12–6   Igneous Rocks—Kilauea Crater, Hawaii

Refer to the map in the appendix, page Ap2–19.

1. Examine the relationship between topographic contours, the faults, and the lavas erupted in 1877, 1919, 1954, and 1959. What type of structural features are these craters?

_____

2. Based on the age of the lavas in the crater, what do you think is the minimum age of Kilauea Crater?

EXERCISE 12–7   Igneous Rocks—Ouray Quadrangle, Colorado

Refer to the map in the appendix, page Ap2–23.

1. Examine the areas underlain by Tsj (the San Juan tuff). At what elevation is the bottom of the tuff exposed on the eastern side of the map? _____
   In the southwestern corner of the map? _____

2. Why isn't the tuff exposed in the valley? _____

3. Describe the contacts of the granodiorite porphyry (gp) in the following places:
   a. Southwestern corner of the map: _____
   b. A mile north of Ouray: _____
   c. Near Twin Peaks (section 36): _____
   d. Northeastern part of map: _____

4. Based on your description, do you think gp was intrusive or extrusive?

_____

_____

_____

5. If it is intrusive, how do you explain the way gp occurs in the southwestern part of the map and near Twin Peaks?

_____

_____

6. A dike labeled "d" occurs south of Ouray. What is the age of this dike?

_____

_____

7. Many mineralized veins (colored red) are present in the area shown on this map. If all of them are about the same age, when did the mineralization take place?

_____

_____

EXERCISE 12–8   Metamorphic Rocks—Greenville Quadrangle, Maine

Refer to the map in the appendix, page Ap2–15.

Note: The rocks exposed in this area are igneous and metamorphic rocks. The red lines with labels S/A, A/B, and B/C are isograds, lines that enable geologists to identify metamorphic zones. C stands for chlorite zone, B is for biotite zone, A is for andalusite-amphibolite zone, and S represents the sillimanite zone. During progressive metamorphism, rocks such as shales are transformed from C to B to A and finally to S as the metamorphism proceeds. In general, rocks pass through each of these stages of metamorphism as temperature and/or pressure increase. The reactions involved are also affected by the amount of water present.

1. Explain why the isogrades form concentric ovals around the outcrops of Dmg and Dpg.

_____

_____

2. Describe the structural features shown on this map.

_____

_____

3. What evidence, if any, suggests that the igneous intrusions took place after the folding of the surrounding metasedimentary rocks?

_____

_____

_____

_____

EXERCISE 14-8 Metamorphic Rocks—Greenville Quadrangle

Name

Refer to the map in the appendix page ...

Selected References

Avery, T. E., and G. L. Berlin. 1992. *Fundamentals of Remote Sensing and Airphoto Interpretation,* 5th ed. New York: Macmillan Publishing Company.

Barnes, J. W. 1981. *Basic Geological Mapping.* Milton Keyes, England: The Open University Press.

Curry, B. B., R. C. Berg, and R. C. Vaiden. 1998. Geologic Mapping for Environmental Planning. McHenry County, Illinois: Illinois State Geological Survey Circular 559.

Gennison, G. M. 1985. *An Introduction to Geological Structures and Maps.* London: Edward Arnold.

Butler, B. C. M., and J. D. Bell. 1988. *Interpretation of Geological Maps.* Burnt Mill, England: Longman Scientific and Technical.

Compton, R. R. 1985. *Geology in the Field.* New York: John Wiley and Sons.

Dennison, J. M. 1968. *Analysis of Geologic Structures.* New York: W. W. Norton and Company, Inc.

Hamblin, W. K. 1980. *Atlas of Stereoscopic Aerial Photographs and Landsat Imagery of North America.* Minneapolis, Minnesota: Tasa Publishing Company.

Lisle, R. J. 1988. *Geological Structures and Maps.* Oxford: Pergamon Press.

Maltman, Alex. 1990. *Geological Maps: An Introduction.* New York: Van Nostrand-Reinhold.

Marshak, Stephen, and Gautam Mitra. 1988. *Basic Methods of Structural Geology.* Englewood Cliffs, NJ: Prentice Hall.

Moseley, F. 1981. *Methods in Field Geology.* Oxford and San Francisco: W. H. Freeman and Company.

Moseley, F. 1979. *Advanced Geological Map Interpretation.* London: Edward Arnold.

Muehrcke, P. C., and J. O. Muehrcke. 1998. *Map Use.* Madison, Wisconsin: JP Publications.

Passcfhier, C. W., J. S. Myers, and A. Kroner. 1990. *Field Geology of High-Grade Gneiss Terrains.* Berlin: Springer-Verlag.

Roberts, J. L. 1982. *Introduction to Geological Maps and Structures.* Oxford: Pergamon Press.

Tucker, M. E. 1982. *The Field Description of Sedimentary Rocks.* Melton Keyes, England: The Open University Press.

Weijermars, R. 1997. *Structural Geology and Map Interpretation.* Amsterdam: Alboran Science Publishing Ltd.

Woodward, N. B., S. E. Boyer, and John Suppe. 1990. *Balanced Geological Cross-sections.* American Geophysical Union.

Index

A P P E N D I X   I

Safety in the Field

You will encounter hazards in every field area. Some hazards are related to weather conditions; others to animals, plants, or insects that may be present. Always seek the advice of someone who is familiar with conditions in the area where you plan to work.

General Safety Guidelines

Wear appropriate clothing.   Footwear is especially important. If you are working on steep slopes, wear boots or shoes with soles that provide good traction. Avoid turned ankles and sprains by wearing boots or shoes that are designed for the type of terrain you will walk across. Carry extra socks. Some people avoid blisters by wearing two pairs of socks. Know the range of temperatures you are likely to encounter, and select suitable clothing. Avoid sunburn by wearing a hat and/or by using sunscreen.

Watch for road hazards.   Be especially careful when you are examining rock outcrops along highways. Stay off the pavement, do not back out on the road without looking both ways, and wear brightly colored clothing or vest. Do not park a vehicle with tires on the pavement.

Avoid dangerous slopes.   Avoid standing at the top or base of a cliff or quarry wall. Rock falls are especially common in the early spring, but people who are upslope may dislodge rocks, and you may trigger a rock fall by hammering at the base of unstable slopes.

Know how to use a geologic hammer.   Never use the pick end of a geologic hammer to break rocks. The pick is suitable for prying rocks loose and for digging into unconsolidated material, but if you hit the tip end of the pick against a hard rock, the tip may break off and become a missile. Wear safety glasses if you are breaking hard, brittle rocks.

Do not go into caves unless experienced spelunkers are with you.   Before you consider going into an undeveloped cave, be familiar with the special rules of safety that apply to caving. A partial list includes (1) go with another person, (2) carry more than one source of light, (3) let someone know where you are going and when you will return, (4) test ropes carefully before using them, and (5) go with someone who is familiar with the cave and with caving techniques.

Watch out for slippery rocks.   Rocks, especially if they are wet and have a thin coating of algae, may be exceptionally slick.

Do not dive into pools.   Never dive into a pool unless you know the water depth and shape of the bottom. Ledges of rock and boulders may be concealed on the bottom.

Poisonous Snakes.   Poisonous North American snakes include three pit vipers—rattlesnakes, copperhead snakes, and water moccasins—all of which have fangs and pits beneath their eyes. The coral snakes, which have neurotoxins, occur only in the southern states. These snakes have red, black, and yellow bands, with the yellow and red bands next to one another. If you are working in a snake-infested area, at least wear long pants and consider wearing leg protection (available through outdoor supply companies).

Poison Ivy, etc.   Learn how to identify poisonous plants that may be in your field area. If you are allergic to poison ivy, wear long pants and long-sleeved shirts. Wash with a strong soap soon after you return from the field. If a rash develops, obtain treatment for it as soon as possible. You can come in contact with the oil from poison ivy plants that is on your clothing.

Insects.   Learn how to identify deer ticks, some of which carry Lyme disease. A large, circular rash or reddish area around a tick bite should be examined by a doctor. Early detection of Lyme disease is important. Insect stings are dangerous to some individuals who have allergic reactions to the stings. If you know you are allergic, carry appropriate medicine with you in the field. If you are stung and break out in hives, experience unusual swelling, or begin to have difficulty breathing, get medical attention immediately.

Allergies and Health Conditions.   If you have serious allergies that could be life threatening or other health conditions such as diabetes, heart problems, or epilepsy, be sure the instructor and others working with you in the field are aware of these conditions. Carry essential medications with you.

Geologic Map Reference Set

Arkansas State Map

Atkinson Creek, Colorado

Davis Mesa, Colorado

Brownwood and Llano, Texas

Calgary, Alberta–British Columbia

Duffield, Virginia

Grand Canyon National Park, Arizona

Greenville, Maine

Jemez Mountains, New Mexico

Kilauea, Hawaii

Mule Mountain, Arizona

Pine Mountain, Colorado

Pittsfield, Massachusetts

Ouray, Colorado

Paradise Peak, Nevada

Salem, Kentucky

Smoky Mountains National Park

Smoky Mountains National Park Cross Section

Williamsport, Pennsylvania

ARKANSAS STATE MAP

Geology by B. R. Haley et al.
USGS, 1993

Map Units:

Quarternary	Qal	Alluvium
	Qt	Terrace deposits
Tertiary	Tj	Jackson Group
	Tc	Clairborne Group
	Tw	Midway Group
Cretaceous	Kn	Nacatoch Sand
	Km	Marlbrook Marl
	Ko	Ozan Formation
	Kb	Brownstown Marl
	Kto	Tokio Formation
	Kt	Trinity Group
	Kdi	Trinity Group-Dierks Limstone
Pennsylvanian	ℙa	Atoka Formation undivided
	ℙau	Upper part
	ℙam	Middle part
	ℙal	Lower part
	Pjv	Johns Valley Shale
	Pj	Jackfork Sandstone
Mississippian	Ms	Stanley Shale
Mississippian & Devonian	MDa	Arkansas Novculite
Silurian	Smb	Missouri Mountain Shale and Blaylock Sandstone
Ordovician	Opb	Polk Creek Shale
	Ow	Womble Shale
	Ob	Blakely Sandstone
	Om	Mazarn Shale
	Ocm	Crystal Mountain Sandstone

Scale 1:500,000

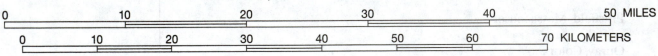

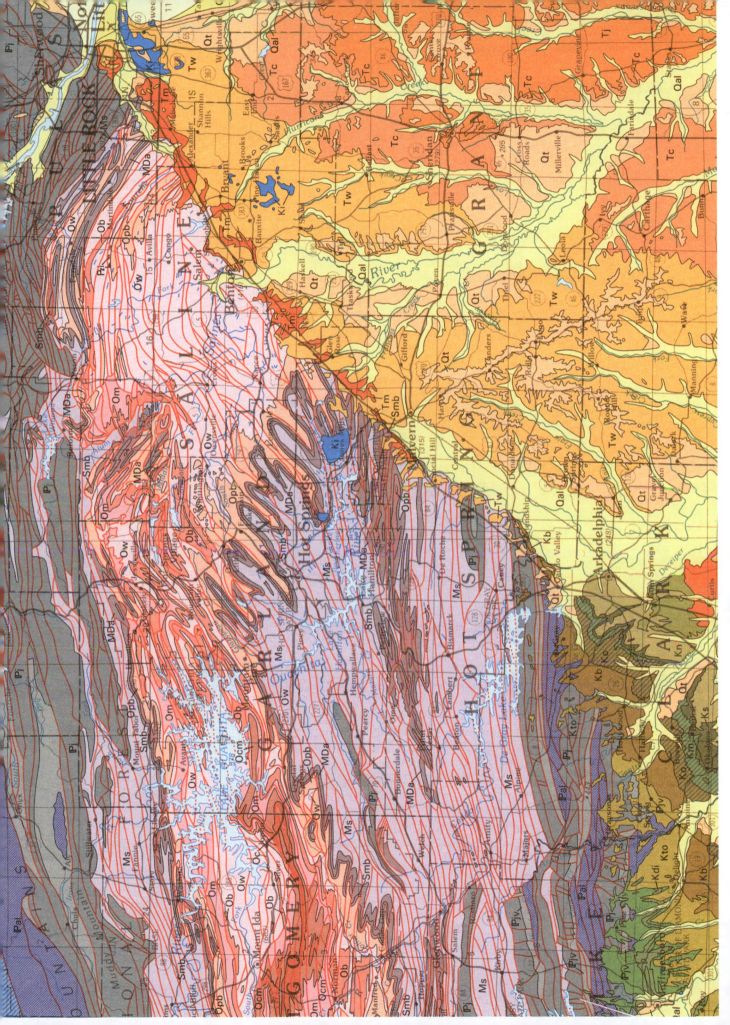

ATKINSON CREEK QUADRANGLE, COLORADO

Geology by E. J. McKay
USGS Map GQ-57, 1955

Topographic contour interval is 20 feet.
Structure contours are drawn on top of the Entrada Formation.
Structure contour interval is 100 feet.

Map Units:

Quaternary	Qal	Alluvium
Cretaceous	Kd	Dakota Formation
	Kbc	Burro Canyon Formation
Jurassic	Jmb	Morrison Formation (upper part)
	Jms	Morrison Formation (lower part)
	Js	Summerville Formation
	Jec	Entrada and Carmel Formations
	Jk	Kayenta Formation
	Jw	Wingate Formation
Triassic	Ŧc	Chinle Formation
Precambrian	pϵ	Schists and gneisses

Scale 1:24,000

```
1            ½            0                         1  MILE
 ▭▭▭▭▭▭▭▭▭▭▭▭▭▭▭▭▭▭▭▭▭▭▭▭▭▭▭▭▭▭▭▭▭▭▭▭▭▭
           1       .5       0              1  KILOMETER
        ▭▭▭▭▭▭▭▭▭▭▭▭▭▭▭▭▭▭▭▭▭▭▭▭▭▭
```

DAVIS MESA QUADRANGLE, COLORADO

Geology by F. W. Cater, Jr.
USGS Map GQ-71, 1955

Topographic contour interval is 20 feet.
Structure contours are drawn on top of the Entrada Formation.
Structure contour interval is 100 feet. (*Note:* elevation of the Entrada
decreases to the northeast.)

Map Units:

Quaternary	Qal	Alluvium
Jurassic	Jmb/Jms	Morrison Formation
	Js	Summerville Formation
	Jec	Entrada Formation
	Jn	Navajo Sandstone
	Jk	Kayenta Formation
	Jw	Wingate Sandstone
Triassic	Ŧc	Chinle Formation
	Ŧmu/Ŧmn/Ŧmb	Moenkopi Formation
Permian	Pc	Cutler Formation

Scale 1:24,000

```
1            ½            0                         1  MILE
 ▭▭▭▭▭▭▭▭▭▭▭▭▭▭▭▭▭▭▭▭▭▭▭▭▭▭▭▭▭▭▭▭▭▭▭▭▭▭
           1       .5       0              1  KILOMETER
        ▭▭▭▭▭▭▭▭▭▭▭▭▭▭▭▭▭▭▭▭▭▭▭▭▭▭
```

BROWNWOOD AND LLANO SHEETS, GEOLOGIC ATLAS OF TEXAS

Project Director, V. E. Barnes
The University of Texas at Austin
Bureau of Economic Geology, 1986
(Reproduced with permission from the Bureau of Economic Geology, The
University of Texas at Austin.)

Map Units:

Qal	Quarternary	Alluvium
Qt	''	Fluviatile terrace deposits
Qu	''	Surficial deposits undivided
Kft	Cretaceous	Fort Terrett Formation
Ka	''	Antlers Sand
Kh	''	Hensell Sand
Ptg	Pennsylvanian	Thrifty and Graham Formations undivided
Phc	''	Home Creek Limestone
Pcc	''	Colony Creek Shale
Ppl	''	Placid Shale
Pw	''	Winchell Limestone
Pab	''	Adams Branch Limestone
Psw	''	Smithwick Shale
Pmf	''	Marble Falls Formation
MD	Mississippian & Devonian	Barnett Fm., Chapel Limstone and Houy Fm. undivided
Oh	Ordovician	Honeycut Formation
Og	''	Gorman Formation
Ot	''	Tanyard Formation
€ws	Cambrian	Wilberns Formation
€wpp	''	''
€wmw	''	''
€rlc	''	Riley Formation
€rh	''	''
p€tm	Precambrian	Town Mountain Granite
p€ps	''	Packsaddle Sch.
p€lc	''	Lost Creek Gneiss
p€vs	''	Valley Spring Gneiss

Scale 1:250,000

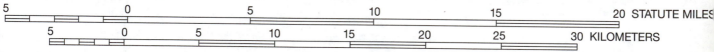

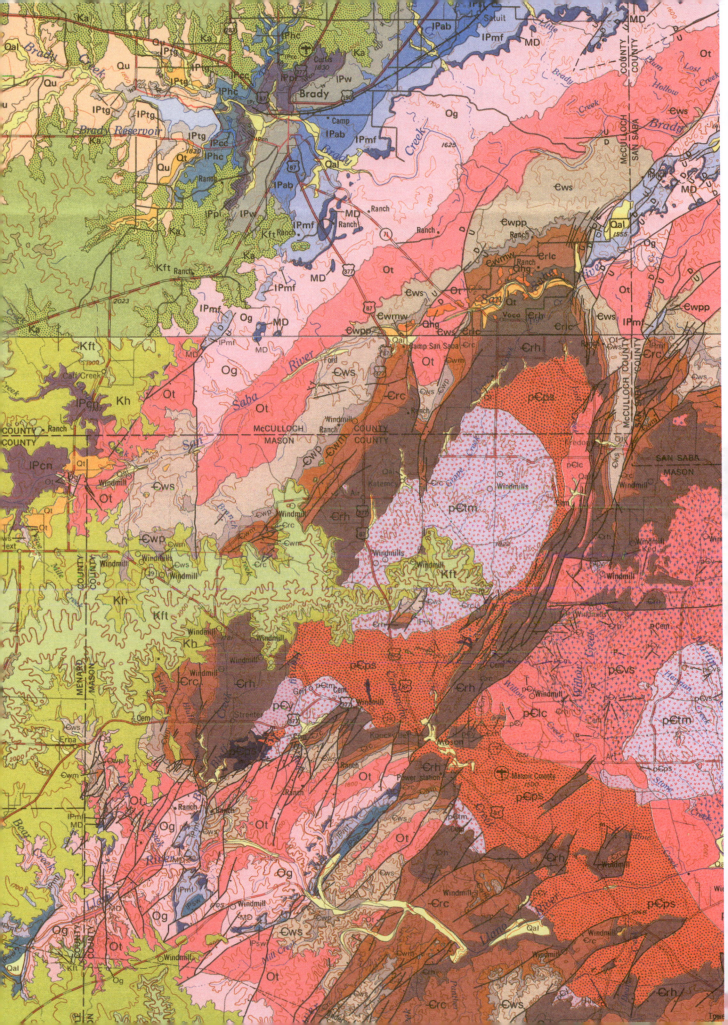

CALGARY, ALBERTA–BRITISH COLUMBIA

Geology by N. C. Ollerenshaw
Geological Survey of Canada, Map 1457A, 1978
(Reproduced with permission from the Geologic Survey of Canada, Natural Resources Canada.)

Note: The Great Plains of Alberta lie east of this map area. The town of Big Prairie is in the plains. Deformation associated with the Canadian Rockies dies out rapidly to the east.

Map Units:

Tertiary	Tpa	Paskapoo Formation
Cretaceous	Kbzu	Brazeau Formation (upper part)
	Kbzl	Brazeau Formation (lower part)
	Kwp	Wapiabi Formation
	Kbkc	Cardium and Blackstone Formations
	Kbmc	Beaver Mines and Mill Creek Formations
	Klbl	Cadomin Formation and Lower Blairmore
Jurassic & Cretaceous	JKk	Kootenay Formation
	Jf	Fernie Formation
Triassic	℞s	Spray River Group
Mississippian, Pennsylvanian, & Permian	M℞PPu	Rundle Group
	Mlv	Livingstone, Pekisko, Shunda Turner Valley Formations
	Mbf	Exshaw and Banff Formations
Devonian	Dpa	Palliser and Alexo/Sasenach Formations
	Df	Cairn, Flume, Southesk, and Mount Hawk Formations
Cambrian	€elx	Eldon Formation to Lynx Group
	€epa	Eldon, Pika, and Arctomoys Formations
	€wcs	Stephen Formations
	€g	Gog Group
Proterozoic	Pmi	Mette Group

Scale 1:250,000

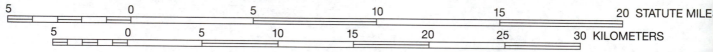

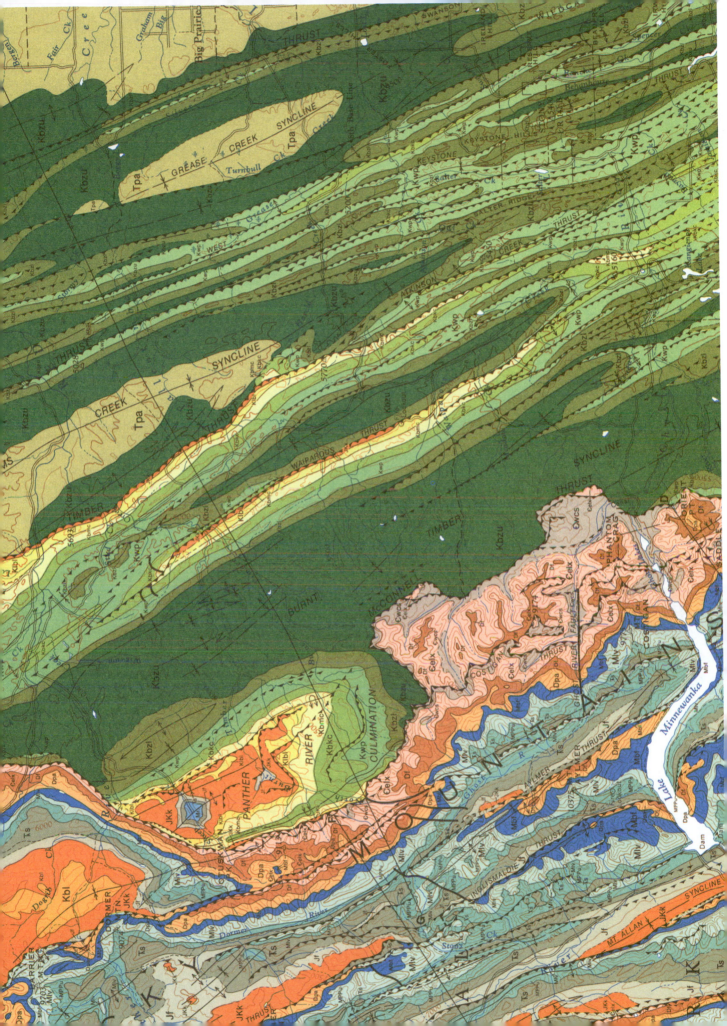

DUFFIELD QUADRANGLE, VIRGINIA

Geology by L. D. Harris and R. L. Miller
USGS Map GQ-111, 1958

Notes: The Hunter Valley Fault lies just southeast of the Clinch River (near the top of the map). The Clinchport Fault is located in the lower right portion of the map. The Red Hill Fault lies north of the Red Hill Church in the upper left portion of the map.

Map Units:

Devonian	Dp	Portage siltstone
	Dg	Genesee Shale
	Dh	Helderberg Formation
Silurian	Sh	Hancock Limestone
	Sct	Clinton Formation
	Sc/Scp/Sch	Clinch Formation
Ordovician	Os	Sequatchie Formation
	Or	Reedsville Formation
	Ot	Trenton Formation
	Oe	Eggleston Formation
	On	Hardy Creek Limestone
	Ob	Ben Hur Limestone
	Owu/Ows	Woodway Limestone
	Olb/Ola	Limestone
	Od	Dot Dolomite
	Oma	Mascot Dolomite
	Okl	Kingsport Formation
	Oc	Chepultepec Formation
Cambrian	€cr	Copper Ridge Dolomite
	€mn	Maynardsville Formation
	€n	Nolichucky Shale
	€m	Maryville Limestone
	€rv	Rogersville Shale
	€rt	Rutledge Limestone
	€p	Pumpkin Valley Shale
	€r	Rome Formation

Scale 1:24,000

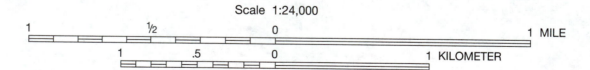

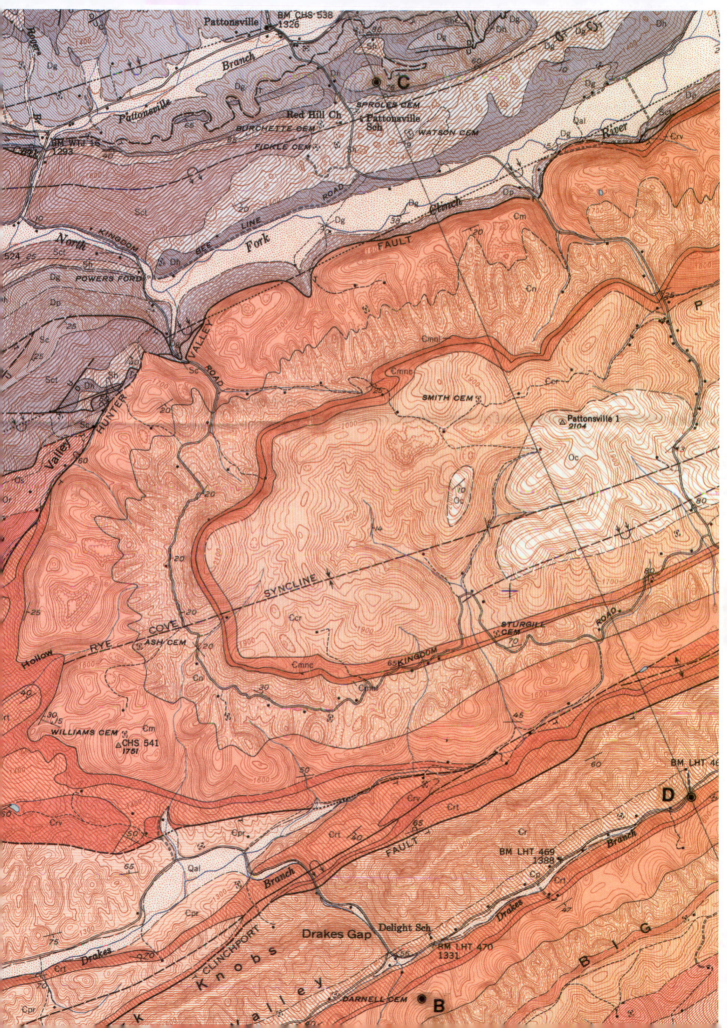

GRAND CANYON NATIONAL PARK, ARIZONA

Grand Canyon Natural History Association and the Museum of Northern
Arizona
(Reproduced with permission from the Grand Canyon Association.)

Map Units:

	s	Slumps, landslides, and rockfalls
	tg	Terrace gravels; loose sand and conglomerate
	r	River deposits; Recent sand, boulders, and mud
Triassic	Ƭs	Chinle Formation; Shinarump Member
	Ƭm	Moenkopi Formation
Permian	Pk	Kaibab Limestone
	Pt	Toroweap Formation
	Pc	Coconino Sandstone
	Ph	Hermit Shale
	Pe	Esplanade Sandstone
Pennsylvanian	ℙs	Wescogam, Manakacha, and Watahomigi Formations undivided
Mississippian	Mr	Redwall Limestone
Devonian	Dtb	Temple Butte Limestone
Cambrian	€m	Muav Limstone
	€ba	Bright Angel Shale
	€t	Tapeats Sandstone
Precambrian	p€k	Kwagunt Formation
	p€g	Galeros Formation
	p€n	Nankoweap Formation
	p€i	Predominantly Diabase intrusives
	p€d	Dox Sandstone
	p€s	Shinumo Quartzite
	p€h	Hakatai Shale
	p€b	Bass Formation
	p€gr1	Zoraster Plutonic Complex
	p€gr2	'' ''
	p€gne	Elves Chasm Gneiss
	p€gnt	Trinity Gneiss
	p€vs	Vishnu Group
	p€va	'' ''
	p€vc	'' ''

Scale 1:62,500

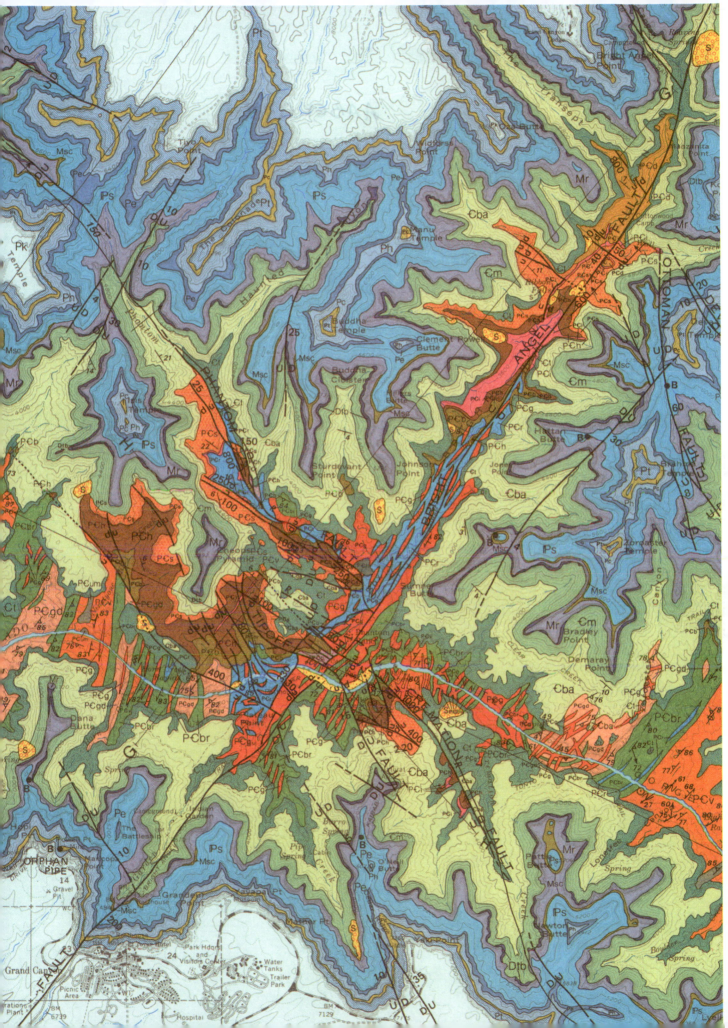

GREENVILLE QUADRANGLE, MAINE

Geology by G. H. Espenshade and E. L. Boudette
USGS Map GQ-330, 1964

Note: The rocks exposed in this area are igneous and metamorphic rocks. The
red lines with labels S/A and B/C are isograds, lines that enable us to identify
metamorphic zones. C stands for chlorite zone, B for biotite zone, A is for
andalusite–amphibolite zone, and S represents the sillimanite zone. During
progressive metamorphism rocks such as shales are transformed from C to B
to A and finally to S as the metamorphism proceeds. In general, rocks pass
through each of these stages of metamorphism as temperature and/or
pressure increase. The reactions involved are also affected by the amount of
water present.

Map Units:

Devonian	Dmg	Quartz monzonite and granodiorite
	Dpg	Pyroxene-Hornblende granodiorite
	Dtng	Troctolite, norite, and gabbro
	Dp	Pelitic rock
	Dps	Pelitic rock
Silurian & Devonian	DSs	Limy Sandstone

Scale 1:24,000

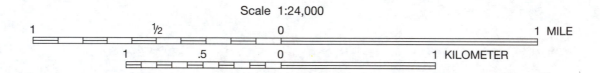

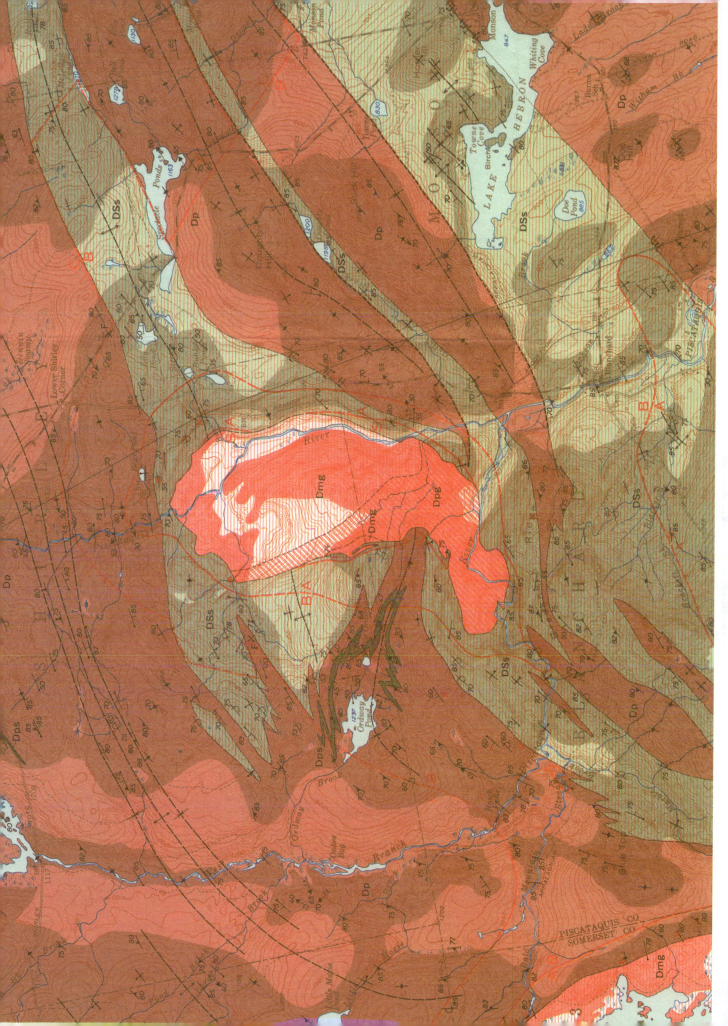

JEMEZ MOUNTAINS, NEW MEXICO

Geology by R. L. Smith, R. A. Bailey, and C. S. Ross
USGS Map I-571, 1970

	Nonvolcanic Sediments		Volcanic Rocks
Quaternary	Qal	Alluvium	Qv = Valles Rhyolite (all symbols starting with Qv)
	Qf	Fan deposits	
	Qls	Landslide deposits	
	Qg	Terrace gravels	
	Qls	Tuffaceous lake deposits	
	Qcf	Caldera fill	
			Qbt, Qbo = Bandelier Tuff
			Qct, ctt = Cerro Rubio Latite
	Qtpg, Qtg,	gravels	Qer = El Rechuelos Rhyolite
	Qtal, Qtp		Tt = Tschicoma Formation
			Tlb = Lobato Basalt

———————————————————— unconformity ————————————————————

Pliocene	Tc	Cochiti Formation	Tbi, Tbf, Tbp = Bearhead Rhyolite
			Tpd, Tpa, Tpb = Paliza Canyon Fm.
			Tcci = Canovas Canyon Rhyolite
			Tcb = Basalt of Chamisa Mesa

———————————————————— unconformity ————————————————————

Miocene	Tsf	Santa Fe Fm. (arkose)	
	Tab	Abiquiu Tuff	
	Tzs	Zia Sand Fm.	
Eocene	Ter	El Rito Fm.	Tvi = rocks of Bland District
	Tg	Galisteo Fm.	

———————————————————— unconformity ————————————————————

Cretaceous	Km	Mancos Shale	
	Kd	Dakota Sandstone	
Jurassic	Ju	undivided	
Triassic	Ŧc	Chinle Fm.	
Permian	Pu	undivided	
Carboniferous	Cu	undivided	

———————————————————— unconformity ————————————————————

Precambrian	pＣu	undivided	

Scale 1:125,000

2   0   2   4   6   8   10 MILES

2   0   2   4   6   8   10 KILOMETERS

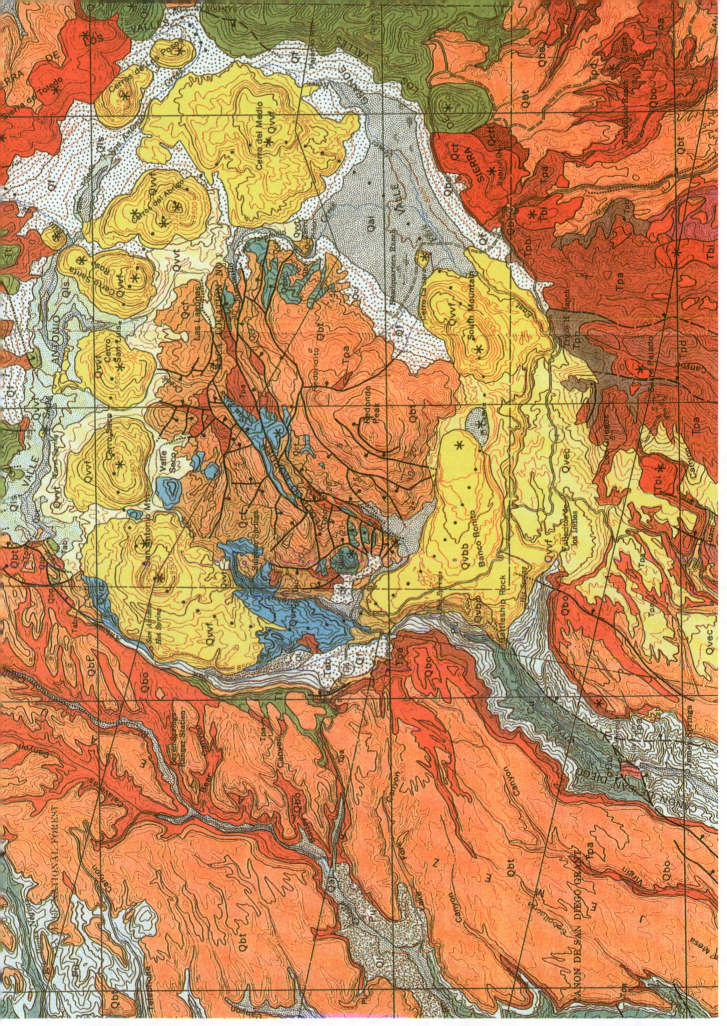

KILAUEA CRATER QUADRANGLE, HAWAII

Geology by D. W. Peterson
USGS Map GQ-667, 1967

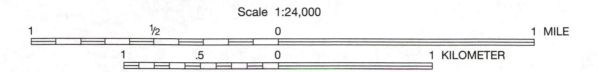

Scale 1:24,000

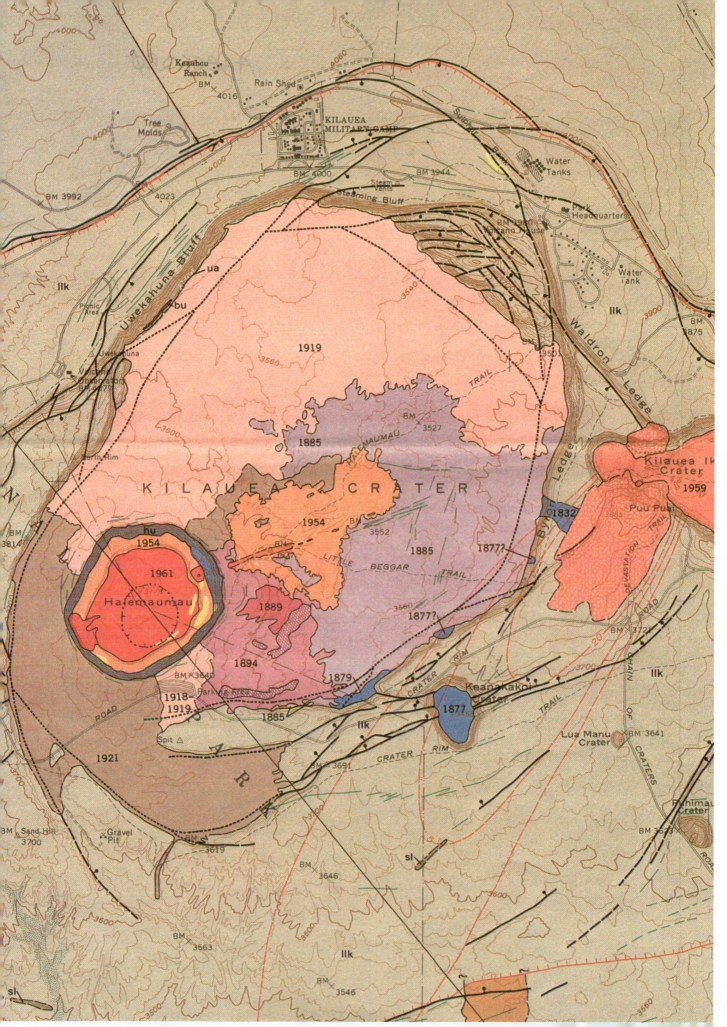

MULE MOUNTAIN QUADRANGLE, ARIZONA

Geology by P. T. Hayes and E. R. Landis
USGS Map I-418

Map Units:

Quaternary	Qg	Pediment, terrace, and fan gravels
Jurassic	Ji	(This intrusion is the pink area in the southwestern corner.)
Cretaceous	Kc	Cintura Formation
	Kmu/Kml	Mural Limestone
	Km	Morita Formation
	Kg	Glance Conglomerate
Devonian	Dm	Martin Limestone
Cambrian	Єb	Bolsa Quartzite
Precambrian	pЄp	Pinal Schist

Scale 1:48,000

1   .5   0   1   2   3   4   5 KILOMETERS

PINE MOUNTAIN QUADRANGLE, COLORADO *(lower left)*

Geology by Fred Cates, Jr.
USGS Map GQ-60, 1955
Structure contours are drawn on the top of the Entrada Formation.
Structure contour interval is 100 feet.

A broad arch, known as the Uncompahgre Uplift, trends northwest–southeast across the northern portion of this map area in western Colorado. A fault extends along the southwestern side of the uplift area.

Map Units:

Quaternary	Qal	Alluvium
	Qfg	Fanglomerate (alluvial fan deposits composed of gravel, conglomerate, sandstone)
Jurassic	Jmb & Jms	Morrison Formation
	Js	Summerville Formation
	Jec	Entrada Formation
	Jk	Kayenta Formation
	Jw	Wingate Sandstone
Triassic	Ťc	Chinle Formation
Permian	Pc	Cutler Formation
Precambrian	pЄ	Gneiss, schist, granite

Scale 1:24,000

1   ½   0   1 MILE

1   .5   0   1 KILOMETER

BEDROCK OF THE PITTSFIELD EAST QUADRANGLE, BERKSHIRE COUNTY, MASSACHUSETTS *(lower right)*

Geology by N. M. Ratcliffe
USGS Map GQ-1574, 1984

Map Units:

Ordovician	Ose	Stockbridge Formation
	Osd	
Cambrian	Єsc	
	Єsb	

Scale 1:24,000

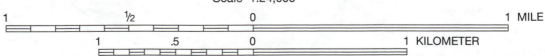

1   ½   0   1 MILE

1   .5   0   1 KILOMETER

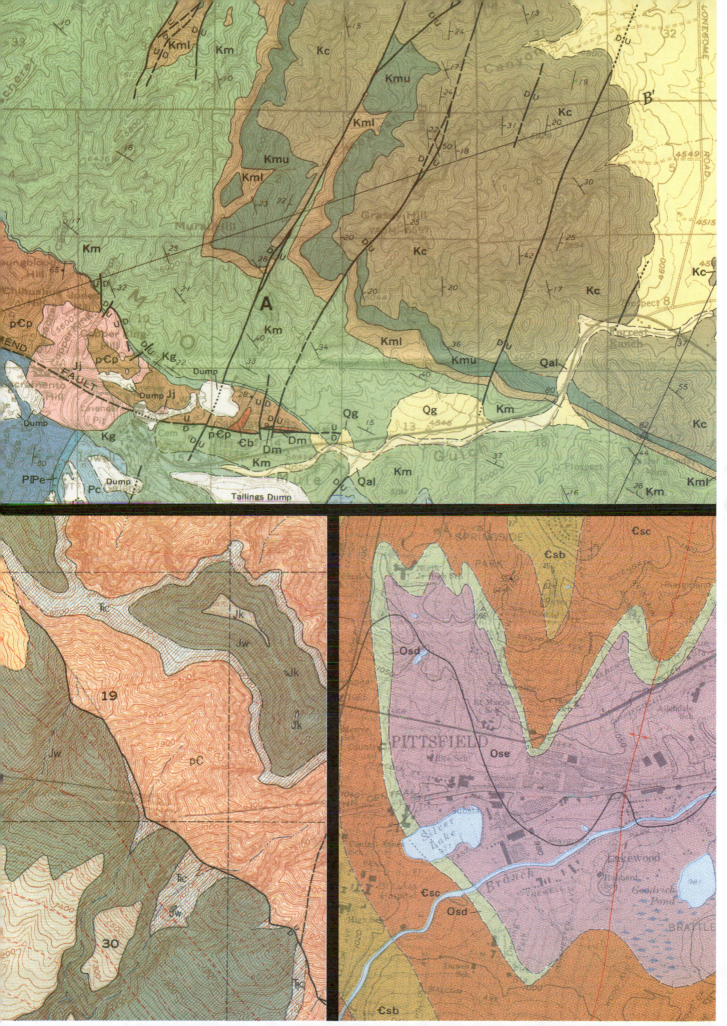

OURAY QUADRANGLE, COLORADO

Geology by R. G. Luedke and W. S. Burbank
USGS Map GQ-152, 1962

Map Units:

Quaternary:	Qu	Undifferentiated deposits
	Qal	Alluvium
	Qf	Alluvial fan and cone deposits
	Qt	Talus
	Qr	Rock glacier deposits
	Qtr	Travertine deposits
	Qs	Landslide deposits
	Qd2	Wisconsin stage glacial deposits
	Qd1	pre-Wisconsin stage glacial deposits
	Qo	Outwash deposits
	Ql	Lake deposits
Tertiary	an	Andesite
	ql	Quartz latite
	Tp	Potosi volcanic series

———————————————————— unconformity

	Ts	Silverton volcanic series

———————————————————— unconformity

	Tsj	San Juan tuff
	fl	Flow in upper part of San Juan tuff

———————————————————— unconformity

	Tt	Telluride conglomerate

———————————————————— (angular) unconformity

	pa	Altered porphyritic rocks
	gp	Granodiorite porphyry
Cretaceous	Km	Mancos shale
	Kd	Dakota sandstone

———————————————————— unconformity

Jurassic	Jm	Morrison formation

———————————————————— unconformity

	Jw	Wanakah formation
	Je	Entrada formation
Triassic	Ʀd	Dolores formation

———————————————————— (angular) unconformity

Permian	Pc	Cutler formation
Pennsylvanian	ℙh	Hermosa formation
	ℙm	Molar formation

———————————————————— unconformity

Mississippian	Ml	Leadville limestone
Devonian	Do	Ouray formation
	De	Elbert formation

———————————————————— (angular) unconformity

Precambrian	p€us	Uncompahgre formation

Scale 1:24,000

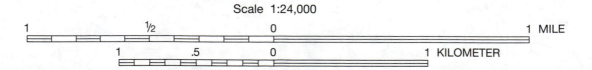

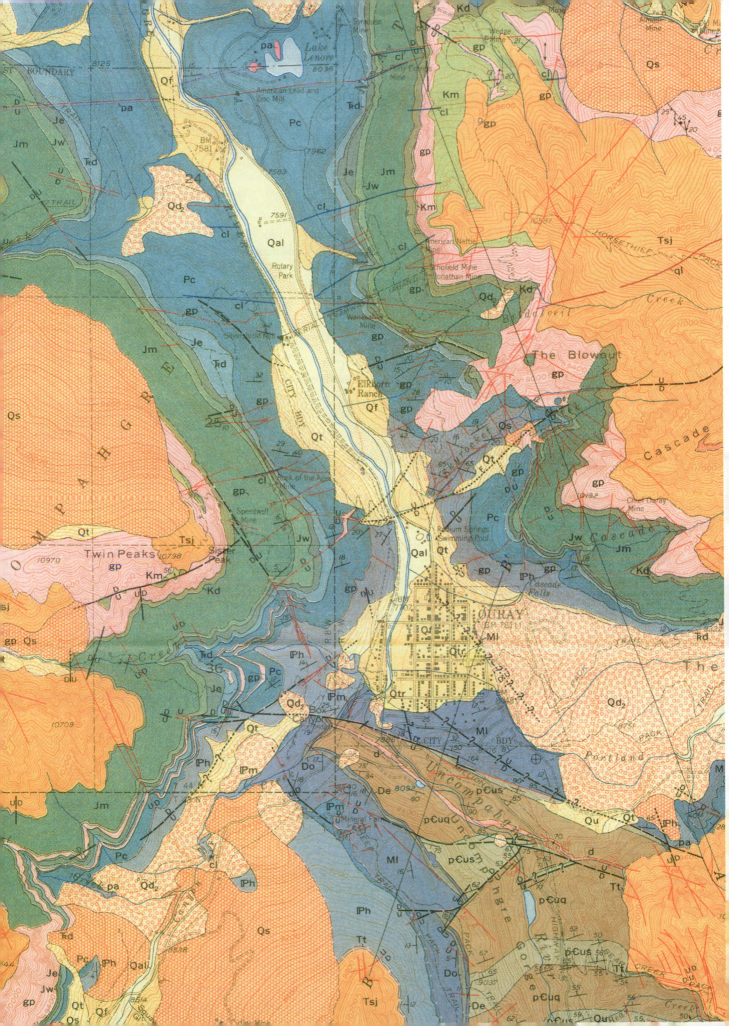

PARADISE PEAK QUADRANGLE, NEVADA

Geology by C. J. Vitaliano and Eugene Callaghan
USGS Map GQ-250, 1963

Map Units:

Quaternary	Qal	Alluvium
	QTi	Intrusive rocks
Tertiary	Tvu	Volcanic rocks (upper sequence)
	Tul	Volcanic rocks (lower sequence)
	gr (red)	Dikes, sills, and plugs of andesitic and rhyolite
	gro	Porphyritic granite
	gry	Younger granite
	gd	Granodiorite
	di	Diorite
	mz	Monzonite
Jurassic	Jd	Dunlap Formation
Jurassic/Triassic	JℝΞu	Limestone, dolomite, siltstone
	JℝΞc	Unit C
	JℝΞb	Unit B
	JℝΞa	Unit A
Triassic	ℝΞldu	Dolomite
	ℝΞldl	Dolomite and limestone
	ℝΞls	Limestone
	ℝΞll	Limestone
Permian	Pq	Quartzite
	Pg	Greenstone

Scale 1:62,500

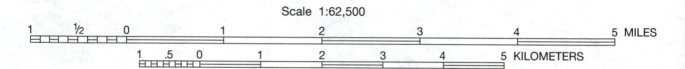

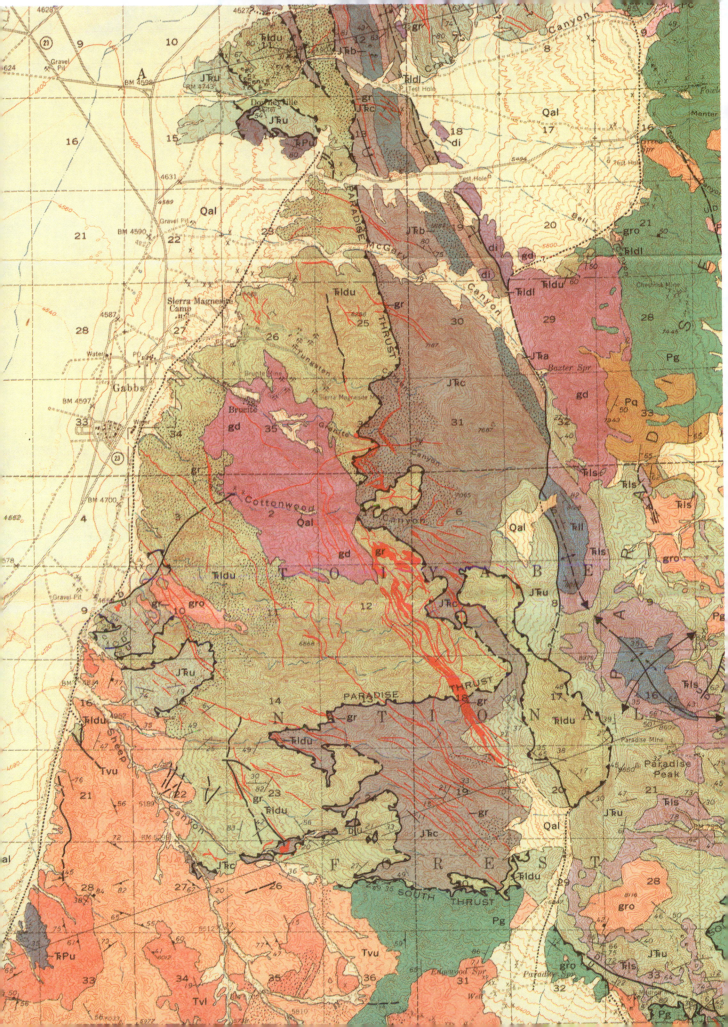

SALEM QUADRANGLE, KENTUCKY

Geology by R. D. Trace
USGS Map GQ-206, 1962

Structure contours are drawn on the base of the Bethel Sandstone.
Hachures indicate closed basin. The structure contour interval is 50 feet.

Map Units:

Quaternary	Qal	Alluvium
Pennsylvanian	Pca	Caseyville Formation
Mississippian	Mkc	Kinkaid Limestone–Clore Limestone
	Mpt	Palestine Sandstone
	Mme	Menard Limestone
	Mwv	Waltersburg Sandstone
	Mts	Tar Springs Sandstone
	Mgd	Glen Dean Limestone
	Mh	Hardinsburg Sandstone
	Mgo	Golconda Formation
	Mhg	Hardinsburg Sandstone and Golconda Formations
	Mcb	Cypress Sandstone
	Mre	Renault Formation
	Msg (Msgl, Msgr, & Msgf)	Ste Genevieve Limestone
	MSL "	St. Louis Limestone

Scale 1:24,000

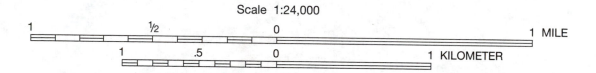